The Car Design Yearbook **2**

Stephen Newbury

The Car Design Yearbook **2**

the definitive guide to new concept
and production cars worldwide

MERRELL
LONDON · NEW YORK

MERRELL

First published 2003 by Merrell Publishers Limited

Head office
42 Southwark Street
London SE1 1UN
Telephone +44 (0)20 7403 2047
E-mail mail@merrellpublishers.com

New York office
49 West 24th Street
New York, NY 10010
Telephone 212 929 8344
E-mail info@merrellpublishersusa.com

www.merrellpublishers.com

Publisher Hugh Merrell
Editorial Director Julian Honer
US Director Joan Brookbank
Sales and Marketing Director Emilie Nangle
Managing Editor Anthea Snow
Editor Sam Wythe
Design Manager Kate Ward
Production Manager Michelle Draycott
Editorial and Production Assistant Emily Sanders

British Library Cataloguing-in-Publication data:
Newbury, Stephen
The car design yearbook 2 : the definitive guide
to new concept and production cars worldwide
1.Automobiles 2.Automobiles – Design
I.Title
629.2'22

ISBN 1 85894 195 4

Consultant editor: Giles Chapman
Edited by Richard Dawes with Laura Hicks
Designed by Kate Ward
Layout by Marit Münzberg
Printed and bound in China

Frontispiece: Jaguar XJ
Pages 4–5: Saab 9-3
Pages 8–9: Subaru B11S Coupé
Pages 18–19: Porsche Cayenne

Contents

Trends, Highlights, Predictions

Since *The Car Design Yearbook* was first published, in September 2002, many of the models it featured as concepts have gone into mass production and many more are nearing their retail debut. In *The Car Design Yearbook 2* we examine all the new production and concept cars unveiled in the intervening twelve months. It's quite a range, from commercially important replacements for established best-sellers up (or down, depending on your viewpoint) to exotic sports cars and wacky showpieces.

The book's intrinsic value is that it is published annually. As you collect each edition it builds into a uniquely comprehensive record that charts automotive design development. In a few years you can look back at how global car design has evolved and see which new ideas have become commonplace.

Once again we profile some important car designers: the work of J. Mays of Ford, Patrick le Quément of Renault and Chris Bangle of BMW is chronicled and analysed. And, in addition, we've provided two special feature sections that will interest the car-design enthusiast and professional: the first is about the objective of concept cars, the second covers the current trend for super-luxury cars. You'll find both of these at the back of the book.

There has been one seismic change in the car industry over the past twelve months. The death of Giovanni Agnelli, head of the Fiat Group, in January 2003 came just as a storm was breaking over the future of the Italian empire started by his grandfather in 1899. The continuing slump in Fiat sales in a viciously competitive market, and the contracted option of selling out to General Motors in 2004, caused a battle in Italy between the government, the Agnelli family and the banks that have propped up Fiat for several years. As Fiat is custodian of the Lancia, Alfa Romeo, Ferrari and Maserati marques, there are rich pickings here for foreign car companies looking for a sporty presence in Europe. Already the vultures are circling. Any break-up of Fiat's embattled empire could see these marques fall into foreign hands – an embarrassing and frustrating scenario for the Italian government.

Designers in the car industry shift from continent to continent, adapting their skills to specific marques and markets. Still, there are very visible differences between the major car-design communities in Europe,

The new Honda Accord is just as inoffensive as its predecessor and yet is set to become another best-seller for the Japanese car giant.

America and Japan. Even minimally informed observers can probably guess where a car is likely to have been designed just by looking at it; but to describe what it is about its 'DNA' that denotes that visual origin is more challenging.

The USA has a long tradition of building large cars and trucks that usually require large-capacity engines to power them. In the country that has the world's biggest domestic market and also some of the tightest speed and environmental restrictions anywhere, these have taken a different evolutionary path from cars in other parts of the world.

From a European or Japanese viewpoint, many American cars lack design subtlety. Then again, to look bold from every angle is perceived as a plus by the US auto executives, who need to keep their products selling. And, don't forget, Detroit has an enviable record as a trend-setting styling innovator, from the Chrysler Airflow and Lincoln Zephyr in the 1930s through the 1950s Studebakers, the 1960 Chevrolet Corvair, 1964 Ford Mustang, 1967 Oldsmobile Toronado and 1983 Chrysler Voyager to the 'cab-forward' stance pioneered by Chrysler saloons in the early 1990s.

Nonetheless, the grilles on American cars continue to be not just striking but also often very dominating. Strong chrome slats are used to emphasize the width, height or power of the car – often all three. As a design starting point this then necessitates dramatic surfaces, which can sometimes make American cars appear overbearing. Surface geometry and lines are not always harmonious, the focus being on large, powerful-looking wheels or an overall impression of toughness and protection. American car design sees no need to shrink from giving people what they want.

At the other end of the scale, Japanese-designed cars conform to a less striking proportion and design language – they seek never to offend. Most Japanese cars are aimed at buyers concerned principally with value and reliability, rather than a look-at-me design statement. This emphasis tends to be expressed in boxy proportions with an athletic flavour, straight feature lines and tight surfaces.

The Japanese have realized there is huge market share to be had with these more mundane-looking models, pitched at buyers who regard cars as consumer durables and not socio-economic iconography. The best example is the archly conventional Honda Accord. Eight million have been sold worldwide since its first, modest appearance in 1976. This is not to say Honda uses dumb design – quite the contrary: it designs cars that will sell to the masses and ensure long-term success in business.

European manufacturers sometimes take a more intellectual approach to design, and there are interesting differences – noticeable, country-specific design traits – between the member countries. Italy possesses emotive brands like Ferrari and Alfa Romeo that use carefully crafted and sporty lines to generate emotion and excitement at the very first glance. The subtlety of these surfaces and lines has been refined over decades by Fiat and Italian design houses like Pininfarina, Bertone and Italdesign – companies that really progress car design and strive to bring new models to market. Recent examples that break the mould include the Fiat Multipla, the Alfa Romeo 147 and a whole raft of brilliant concept cars. How successful Fiat would have become if its Italian-inspired designs had been matched by Japanese standards of engineering quality is worth pondering.

In Germany, the combination of sportiness and conservatism is a constant theme. Volkswagen dominates the market with the hugely practical Golf – well built, square and solid-looking. Other brands, such as BMW, have perfected their DNA over the decades, an evolution that has created the ultimate in sporty drivers' cars, while Porsche has built itself up through success in motorsport. Teutonic interiors and absolute functionality take preference over emotional aesthetics, resulting in an engineered look. The new Porsche Cayenne struggles to combine its SUV proportions with Porsche's sports-car design DNA, but owing to its functionality, build quality and brand strength this car is guaranteed to be a big success.

In France, outstanding design from Renault, Peugeot and Citroën challenges rival manufacturers around the world with arresting shapes and clever concepts, happily proving that even in this time of over-capacity it's possible to make a profit without being a multi-national company. The Renault–Nissan alliance is fascinating. As economies of scale kick in and a new Nissan corporate image is carried through to the whole range, this partnership will be a great force in Europe. The C-Airdream concept from Citroën and the new Renault Espace demonstrate that the French approach interior design differently

Opposite and above
Many car manufacturers are trying to exploit new vehicle niches to capitalize on the heavy overall investments they need to make. The new Cayenne from Porsche, the result of a joint venture with Volkswagen, is one example that also has the advantage of carrying an extremely strong marque badge.

Acura TSX
Alfa Romeo Kamal
Aston Martin AMV8 Vantage
Audi A3
Audi A8
Audi Nuvolari
Audi Pikes Peak
Bentley Continental GT
Bertone Birusa
BMW xActivity
BMW Z4
Buick Centieme
Cadillac Sixteen
Cadillac SRX
Chevrolet Aveo
Chevrolet Cheyenne
Chevrolet Colorado
Chevrolet Equinox
Chevrolet Malibu
Chevrolet SS
Chrysler Airflite
Chrysler California Cruiser
Chrysler Pacifica
Citroën C-Airdream
Daewoo Flex
Daewoo Nubira
Daewoo Scope
Dodge Avenger
Dodge Durango
Dodge Kahuna
Dodge Magnum SRT-8
Ferrari Enzo
Fiat Gingo
Fiat Idea
Fiat Marrakech
Fiat Simba
Ford 427

Ford F-150
Ford Faction
Ford Focus C-Max
Ford Freestar
Ford Freestyle FX
Ford Model U
Ford Mustang GT
Ford Streetka
GM Hy-wire
Honda Accord
Honda Element
Hyundai HIC
Infiniti FX45
Infiniti M45
Infiniti Triant
Invicta S1
Italdesign Moray
Jaguar XJ
Kia KCD-1 Slice
Kia KCV-II
Lamborghini Gallardo
Lancia Ypsilon
Lexus RX330
Lincoln Aviator
Lincoln Navicross
Matra P75
Maybach
Mazda2
Mazda MX Sportif
Mazda Washu
MCC Smart Roadster/Roadster-Coupé
Mercury Messenger
Mercury Monterey
MG XPower SV
Mitsubishi Endeavor
Mitsubishi Tarmac Spyder
Nissan Evalia

Nissan Maxima
Nissan Micra
Nissan Murano
Nissan Quest
Nissan Titan
Opel GTC Genève
Opel/Vauxhall Meriva
Opel/Vauxhall Signum
Peugeot H$_2$O
Peugeot Hoggar
Peugeot Sésame
Pininfarina Enjoy
Pontiac G6
Pontiac Grand Prix
Porsche Cayenne
Renault Ellypse
Renault Espace
Renault Mégane
Renault Scénic 2
Rinspeed Bedouin
Rolls-Royce Phantom
Saab 9-3
Saturn Ion
Scion xA
Scion xB
Seat Cordoba
Sivax Streetster Kira
Subaru B11S Coupé
Suzuki Forenza
Suzuki Verona
Toyota Land Cruiser
Toyota Sienna
TVR T350
Volkswagen L1
Volkswagen Touareg
Volvo VCC

A–Z of New Models

Acura TSX

Design	Honda Motor Corporation
Engine	2.4 in-line 4
Power	149 kW (200 bhp) @ 6800 rpm
Torque	225 Nm (166 lb. ft.) @ 4500 rpm
Gearbox	6-speed manual
Front suspension	Double wishbone
Rear suspension	Multi-link
Brakes front/rear	Discs/discs
Front tyres	P215/50R17
Rear tyres	P215/50R17
Length	4657 mm (183.3 in.)
Width	1762 mm (69.4 in.)
Height	1456 mm (57.3 in.)
Fuel consumption	9.4 ltr/100 km (30 mpg)

The all-new Acura TSX sports saloon went on sale in the US during the spring of 2003. This new model, with its sharply wedged design, is aimed directly at grabbing market share from European competitors such as BMW's 3-series and Volvo's S60.

The TSX will sit in Acura's line-up between the RSX sports coupé and Acura's top-selling model, the 3.2 TL executive saloon.

With a distinctive 'V'-shaped grille and trapezoidal headlights, the look is characterful and sporty. In fact, the silhouette, together with some of the feature lines, could have been lifted straight from the Alfa Romeo 156, a car that has enjoyed great success in Europe among design-conscious buyers. Significantly, the 156 has also eaten into BMW territory.

At the back, the wide C-pillars wrap around to a rear screen that falls down to the short boot lid, in that sporty, typically BMW 3-Series way. The boot drops away vertically, past the narrow rear lights and on to a simple rear bumper that houses twin chrome tail pipes.

The cockpit matches the outside in its sportiness, and also uses good-quality materials. However, the combination of perforated leather on the sports seats with black leather, wood and aluminium detailing gives a look that is now overused and tired; the aluminium, in particular, is stark among the other, softer shades.

Although the instruments are clearly visible through the sporty, three-spoke steering wheel, the interior doesn't have the emotional flair of the exterior, even though it's obviously well appointed. Few marks here, then, for design innovation.

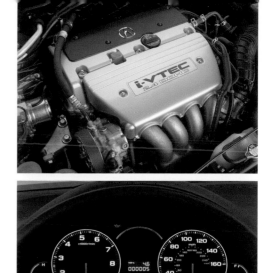

Alfa Romeo Kamal

At the Geneva Motor Show in 2003, Alfa Romeo showed the Kamal, designed at the Arese Style Centre in Italy. Not a working car, this is an exterior model of an envisaged Alfa SUV with a design strongly linked to the production 147 but with more voluptuous surfaces and a more purposeful stance. The SUV niche is one that most other manufacturers have been attacking with gusto of late.

The name 'Kamal' comes from the ancient Sanskrit language, where it means 'red', a colour long associated with the Alfa Romeo insignia. In Arabic the same word means 'perfection' or the 'synthesis of opposites'. All these epithets are fine as associations, the Fiat-controlled brand has decided, and indeed anything that can boost the appeal of Fiat-made products right now is welcome, no matter how at odds it is with company tradition, for the fact is, Alfa Romeo has never built an SUV before.

The front is in traditional Alfa Romeo style, with the design lines all drawing the eye to the point of the centre grille. The strong openings either side of the grille, reminiscent of those of the BMW Z8, give a very powerful, sporty impression. The bulges in the wheel arches are highlighted by the sinuous waistline, which droops over the front door before rising up to finish at the rear light.

A car like this could achieve good sales in the current market, and an Alfa Romeo would make an intriguing alternative to the mainly German, Japanese and British options on offer in Europe. The model could share many components with the 147 and 156, but Alfa's typical fragility – fine in sports and racing cars but no good if you're stuck in the middle of nowhere – needs to be banished if massive warranty claims are not to be an unhappy by-product of selling a real-life Kamal.

Engine	3.2 V6
Power	186 kW (250 bhp) @ 6200 rpm
Torque	300 Nm (221 lb. ft.) @ 4800 rpm
Gearbox	6-speed manual
Installation	Front-engined/all-wheel drive
Height	1620 mm (63.8 in.)

Aston Martin AMV8 Vantage

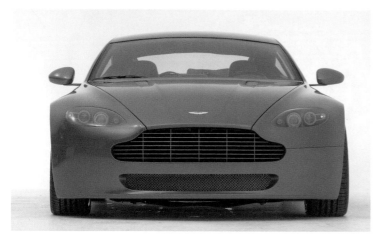

Design	Henrik Fisker
Engine	4.3 V8
Installation	Front mid-engined/rear-wheel drive
Brakes front/rear	Discs/discs
Length	4347 mm (171.1 in.)
Width	1874 mm (73.8 in.)
Height	1298 mm (51.1 in.)
Wheelbase	2600 mm (102.4 in.)
Kerb weight	1500 kg (3307 lb.)

An impressive new model to be launched by Aston Martin in 2005 is seen here in concept form as the AMV8 Vantage. It's intended as a new, small Aston Martin that will compete directly with Porsche – a size of car not made by the British company since the DB2 series back in the 1950s. In addition, this aims to boost Aston Martin's production significantly, and it could turn out to be the most important launch in the company's eighty-nine-year history.

The name Vantage has historic links with Aston Martin; it first appeared on the DB2 in 1950 to denote a high-performance engine option. This time, though, it's used to identify a specific model name. The AMV8 Vantage is a two-seater sports car retaining a front-engined/rear-wheel-drive layout for 50:50 weight distribution. Much of the car's structure will evolve from the top-end Vanquish, which features aluminium and composites. According to Aston Martin, the exterior design is 90% true to the production version, but the interior will be reworked for 2005.

Traditional Aston Martin design cues have been used, including the front grille shape and the side air-intake strakes, as found on all Aston Martins. The AMV8 Vantage is, however, a thoroughly modern interpretation of the company's heritage.

Inside, the designers have – sensibly – specified everything you see and touch as uniquely Aston Martin and not taken from a Ford parts bin. After all, if TVR can do that at half the price, then Aston Martin customers should expect it too.

The interior is finished in an imaginative combination of different leathers and anodized aluminium. Customers will be offered a huge choice of upholstery schemes and the option of different materials on request. A wide range of body colours will also be available.

Even though production numbers will be higher than for existing models such as the DB7, the car should still be pretty exclusive. And that is one characteristic of the British marque that can't be altered too radically.

Aston Martin AMV8 Vantage **27**

Audi A3

Design	Peter Schreyer
Engine	3.2 V6 (2.0 and 2.0 diesel in-line 4 also offered)
Torque	320 Nm (236 lb. ft.)
Gearbox	6-speed manual
Installation	Front-engined/all-wheel drive
Front suspension	MacPherson strut
Rear suspension	4-link
Length	4203 mm (165.5 in.)
Width	1765 mm (69.5 in.)
Height	1421 mm (55.9 in.)
Wheelbase	2578 mm (101.5 in.)
Kerb weight	1348 kg (2972 lb.)

Since its launch in 1996 the Audi A3 has achieved fantastic sales in the premium-hatchback segment by combining a high-quality product with conservative and safe design that appeals strongly to the middle-class market. A bit 'square', some might think, though others would say stylishly timeless.

The new model is identical in its approach – obviously a sensible move, as it would be foolish to scare off a loyal customer base. But there are some changes. In fact, probably the best way to describe the new look is that it has been sharpened up and, as a result, the new A3 is slightly sportier than its predecessor. At the front, the headlights are now clear, with technical-looking reflectors inside them. There are crisper lines in the bonnet and the front bumper now has sharp, vented air intakes with small radii. Along the side there's now a sharp, rising feature line that extends through the fuel filler flap to the rear lights. The rear is constructed mainly of horizontal and vertical lines, and is slightly less curvaceous than in the previous model.

Interior space has been improved, thanks to an increase of 65 mm (2.5 in.) in the wheelbase. The sober interior design is very similar to what was offered last time, made up of taut lines and featuring dark- and light-grey materials.

The new, sharper A3 went on sale during the summer of 2003 and is surely destined to continue the success for Audi in this small segment, one that it helped shape and now utterly dominates. It must be thankful it possesses a design that can mature so assuredly.

Audi A8

Engine	4.2 V8 (3.7 V8 also offered)
Power	246 kW (330 bhp) @ 6500 rpm
Torque	430 Nm (317 lb. ft.) @ 3500 rpm
Gearbox	6-speed Tiptronic
Installation	Front-engined/all-wheel drive
Front suspension	Double-wishbone air suspension
Rear suspension	Trapezoidal-link axle with air suspension
Brakes front/rear	Discs/discs, ESP, ABS, EBD, BA
Front tyres	235/55R17
Rear tyres	235/55R17
Length	5051 mm (198.9 in.)
Width	1894 mm (74.6 in.)
Height	1444 mm (56.9 in.)
Wheelbase	2944 mm (115.9 in.)
Track front/rear	1629/1615 mm (64.1/63.6 in.)
Kerb weight	1780 kg (3924 lb.)
0–100 km/h (62 mph)	6.3 sec
Top speed	250 km/h (155 mph) governed
Fuel consumption	12 ltr/100 km (23.5 mpg)
CO$_2$ emissions	287 g/km

The new A8 is the brand flagship model for Audi. This is a model that is renowned for pioneering new technology: the first A8, introduced in 1994 to replace the slow-selling and rather anonymous Audi V8, was the first luxury car to use aluminium for both frame and body; indeed, it was the first all-aluminium car to go into volume production. This is a weight-saving and environmentally friendly route to executive car-making that has been emulated by Jaguar with its new XJ.

Although the exterior is subtly different from that of the old model, particularly at the rear, the new internal technology is the most interesting aspect of the new A8. The aluminium body is based on a further development of the Audi Space Frame (ASF); Audi claims it is stiffer and more refined.

For the chassis, adaptive air suspension with continuously adjustable damping gives an optimum balance between comfort and handling, depending on the driving mode selected. The driver can choose from one of four pre-defined settings: an automatic mode where the body is at the standard height but lowered by 25 mm (1 in.) if the car is driven at more than 120 km/h (75 mph); a dynamic mode where the car is lowered by 20 mm (0.8 in.) before it sets off and the air suspension operates with firmer springs and a harder damping characteristic; a comfort mode to help the car glide smoothly over bumpy surfaces; and a 'lift' mode, designed for driving on uneven terrain, which can be activated below 80 km/h (50 mph) to raise ground clearance by 25 mm (1 in.) with the balanced damping characteristics of the automatic mode.

Other electronic systems fitted include the electro-mechanical parking brake and the radar-assisted distance control system known as 'adaptive cruise control'.

The original A8 was regarded as an all-round high performer, albeit one with a less defined character than its rivals. But this new model promises a lot more, and will surely bite much harder into the sales of the new BMW 7 Series.

Audi Nuvolari

The Audi Nuvolari design study is named after the motor-racing driver Tazio Nuvolari, the last man to win a grand prix in an Auto Union car, in Belgrade in 1939. Nuvolari died in 1953 but is still famous for his daring, spectacular driving style and the yellow pullover he always wore at the wheel.

This quattro concept study is a vision for a powerful Gran Turismo car, a two-door two-plus-two coupé with GT proportions. Its design lines are superficially similar to those of the Audi TT, but this GT has much more presence.

The long bonnet, the high waistline and the arched roofline that drops smoothly at the rear dominate its silhouette; the high waistline makes the windows extremely shallow, as in the Audi TT. At the side, though, is a dynamic line in the sill, and higher up is a feature line that clips the edge of the bulging wheel arches housing the massive wheels, which are set with minimum overhang.

Some new technology is featured. The LED headlights have allowed the designers more freedom because they need less installation space than conventional ones. For enhanced personal protection, two cameras check for occupant position in the event that the airbag needs to be inflated. For security, the glove compartment is opened by fingerprint recognition – cute, if a little paranoid.

Inside, leather and aluminium materials are combined: contrasting Stromboli black and Carrara white for the leather, while cool aluminium adds design structure to the dashboard, doors and centre console, all designed to encapsulate GT philosophy.

Audi claims the Nuvolari outlines the future direction for the brand. If this is true, it's an extremely exciting proposition.

Design	Walter De Silva
Engine	5.0 V10 bi-turbo
Power	447 kW (600 bhp)
Torque	750 Nm (553 lb. ft.) @ 2000 rpm
Installation	Front-engined/all-wheel drive
Front suspension	4-link
Rear suspension	Trapezoidal link
Brakes front/rear	Discs/discs
Front tyres	265/720 R 560 PAX
Rear tyres	265/720 R 560 PAX
Length	4800 mm (188.9 in.)
Width	1920 mm (75.6 in.)
Height	1410 mm (55.5 in.)
0–100 km/h (62 mph)	4.1 sec
Top speed	250 km/h (155 mph)

Audi Pikes Peak

Not wanting to be left behind by the SUV bandwagon, Audi has designed this quattro. The Pikes Peak combines massive power with all-terrain capability and MPV versatility. The name comes from the Pike's Peak hill-climb race held in Colorado, USA, every June – an event Audi has won three times.

Externally, the Pikes Peak is an up-to-the-minute interpretation of the Audi Allroad but with dedicated off-road capability and more power. Technical details on the exterior include loop-pattern door handles that extend when proximity sensors detect the key-holder is approaching. The handles are also illuminated, and lights in the exterior mirrors illuminate the area around the doors. The body is set off by huge five-arm, double-spoke wheels.

The interior is also innovative. A four-plus-two configuration allows the Pikes Peak to be adapted from sports car to off-roader to people-carrier. Off-road competence comes from the car's variable-height adaptive air suspension and four-wheel-drive drivetrain. The ride height can be raised for rough terrain but is automatically lowered when the car is back on smooth roads. A glass roof provides plenty of light to the interior, where leather-upholstered seats accommodate up to six people. The interior also includes laser optics, a DVD entertainment system and internet access.

Audi says that while there are currently no plans to put the Pikes Peak into full production, it will study public opinion very closely. Manufacturers always say that. Still, the car would do extremely well in the current market despite a plethora of existing and upcoming competitors.

Design	Walter De Silva
Engine	4.2 V8
Power	373 kW (500 bhp)
Installation	Front-engined/4-wheel drive
Brakes front/rear	Discs/discs
0–100 km/h (62 mph)	4.9 sec

Bentley Continental GT

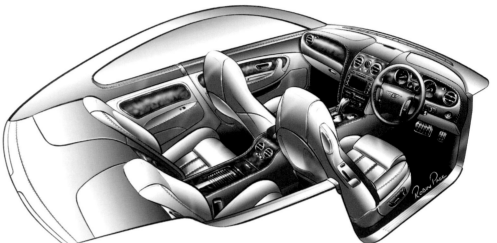

Now under the ownership of the VW Group, Bentley finds itself for the first time since the 1950s (and the genesis of the original R-type Continental) in the enviable position of being allowed to invest heavily in new products. Since 1965 all Bentleys have been based on derivatives and descendants of the Rolls-Royce Silver Shadow platform, and this has hampered handling, performance and general sportiness. The new Continental GT changes all that, and received great acclaim at the 2002 Paris Motor Show.

This long-anticipated new model is based on the Volkswagen Phaeton platform, which also underpins the new Audi A8. It is fitted with a Bentley-developed version of Volkswagen's W12 engine, a massively powerful and smooth unit that is key to the Continental's performance promise. This new sports coupé is not just the fastest model in Bentley's illustrious eighty-three-year history: it also happens to be the fastest four-seater coupé in the world.

The design philosophy behind the Continental GT was to produce a vehicle with true supercar performance that carries four people and their luggage and displays the highest levels of refinement. References are made to the 1950s R-type Continental – in particular the 'suggested' front and rear wings, the gently tapering fastback and the interplay between the side window shape and the tautly gathered waistline. Yet much of the design 'formality' of recent Bentleys has been discarded: obtrusive, 'separate' bumpers are gone; the Bentley grille is a stylized shadow of its former gaunt self; and the seamless, harmonious integration of body panels has an unmistakable aura of current Audi styling.

Inside, the cabin remains true to the marque's values, with chrome, fine woods and leather that adorn the space as in almost every Bentley of the past. The instrument panel now splits at the centre console to distinguish the driver's space from the front passenger's. The British veneer remains, but the Continental GT is vastly more contemporary European in its interior lines.

Crusty Bentley traditionalists will need time to adjust, but, crucially, the new model catapults Bentley into the twenty-first century and is a brand-new basis from which other derivatives can be explored.

BMW Z4

Design	Chris Bangle
Engine	3.0 in-line 6 (2.5 in-line 6 also offered)
Power	170 kW (231 bhp) @ 5900 rpm
Torque	300 Nm (221 lb. ft.) @ 3500 rpm
Gearbox	6-speed manual
Installation	Front-engined/rear-wheel drive
Front suspension	MacPherson strut
Brakes front/rear	Discs/discs
Front tyres	225/45R17
Rear tyres	225/45R17
Length	4091 mm (161.1 in.)
Width	1781 mm (70.1 in.)
Height	1299 mm (51.1 in.)
Wheelbase	2495 mm (98.2 in.)
Kerb weight	1290 kg (2844 lb.)
0–100 km/h (62 mph)	5.9 sec
Top speed	250 km/h (155 mph)
Fuel consumption	9.11 ltr/100 km (31 mpg)
CO$_2$ emissions	221 g/km

The BMW Z4 roadster replaces the Z3, the American-built two-seater sports car launched in 1995 that was widely lambasted by motoring critics for its design and lacklustre performance. This panning, though, was no bar to popularity among buyers whose main criteria were style, prestige and ease of ownership, and now BMW's design chief, Chris Bangle, has righted the entry-level roadster and endowed it with a much purer sports appeal.

Based closely on the style of the CS1 concept car shown at Geneva in 2002, the Z4 dispenses with the two rear seats to give a dedicated two-seater layout and also makes a number of refinements to the exterior surfaces. Gone is the horizontal waistline in favour of a slightly more curvaceous side profile. Added is the wing crease that strikes through a large BMW badge in front of the doors, and the boot lid now has a moulded-in spoiler complete with centrally high-mounted stop-light (the industry abbreviation is CHMSL!) The new Z4 is much more coherent in its design as a muscular car. By comparison, the Z3 had a rather 'masculine' front yet a 'feminine' tail treatment.

The Z4's interior enjoys classic roadster design that is straightforward, clear, modern and elegant, with black and silver as the main colour finishes. The instrument panel is curved towards the inside; the centre console is shaped like a letter 'T' facing towards the instrument panel. Conventionally in a sports car, the speedometer and rev counter dominate the dashboard, with tube-like covers to protect them against distracting light reflection from the windscreen.

The complex exterior surfaces of this car generate unique shadows, depending on the light and the angle from which you view it. Chris Bangle can only be commended for bringing the Z4 to production reality. It looks great, and is no doubt the perfect machine for a weekend jaunt on country lanes.

Buick Centieme

Design	Gary Mack
Engine	3.6 V6 twin-turbo
Power	298 kW (400 bhp)
Torque	543 Nm (400 lb. ft.)
Gearbox	4-speed automatic
Installation	Front-engined/all-wheel drive
Front suspension	MacPherson strut
Rear suspension	Short and long arm
Brakes front/rear	Discs/discs
Front tyres	P275/40R22
Rear tyres	P275/40R22
Length	4786 mm (188.4 in.)
Width	1971 mm (77.6 in.)
Height	1634 mm (64.3 in.)
Wheelbase	3026 mm (119.1 in.)
Track front/rear	1697/1709 mm (66.8/67.3 in.)
Kerb weight	1792 kg (3951 lb.)

Buick launched the Centieme to commemorate its one-hundredth anniversary. The Italian design consultant Bertone built this concept, a luxurious vehicle combining the features of a saloon with those of a typical SUV.

The Centieme seats six in rows of two. It has a low, wide stance, and sports flowing surfaces that are perhaps too curvaceous, feature lines and a classic, if dominating, Buick grille. Combined with a relatively long wheelbase, tight overhangs and a rising daylight opening line, these features give the car a nimble and energetic appearance.

Inside, the luxury is evident, with no expense spared. The seating for the first two rows uses captain's chairs. Armrests are located on the adjacent doors and integrated in the seats for symmetry and greater comfort. Porcelain-coloured leather covers the seats and the lower doors. Dusk-coloured leather on the third-row seating and upper doors contrasts beautifully with the lighter hue of the first and second rows, giving the cabin a beguiling ambience.

Porcelain soft-touch suede is used for the headlining, while the leather floor is custom-made and hints at highly contemporary home decor. The steering wheel, consoles and interior trim are accented in olive ash, and the cluster gauges have aluminium detailing.

The Centieme is designed to celebrate a century of Buick heritage while harnessing contemporary values. Sleek, romantic lines accentuate full sculptural forms to create what might well be the future for Buick designs. But remember: this is a show car, pure and simple.

Cadillac SRX

Design	Kip Wasenko
Engine	4.6 V8
Power	235 kW (315 bhp) @ 6400 rpm
Torque	421 Nm (310 lb. ft.) @ 4400 rpm
Gearbox	5-speed automatic
Installation	Front-engined/rear-wheel drive
Front suspension	Short and long arm
Rear suspension	Multi-link
Brakes front/rear	Discs/discs
Front tyres	P235/60R18
Rear tyres	P255/55R18
Length	4950 mm (194.9 in.)
Width	1844 mm (72.6 in.)
Height	1671 mm (65.8 in.)
Wheelbase	2957 mm (116.4 in.)
Kerb weight	1948 kg (4295 lb.)

The Cadillac SRX clearly reflects Cadillac's new design language as evidenced in several concept cars over the past four years. These include the Evoq in 1999, the Imaj in 2000, the Vizon in 2001 and the Cien supercar in 2002. A shorthand definition of this design language would be: structural and boxy proportions, chiselled forms and straight feature lines.

This approach endows the SRX with a tough-looking stance that will clearly make it stand out of the crowd of other, much softer SUVs – the Infiniti FX45 or the dramatic Mercedes-Benz GST (when it eventually arrives) spring to mind.

The SRX shares some design elements with the CTS production saloon, such as the 'V'-shaped grille and vertical tail lights and headlights – all features that appeared on Cadillacs back in 1965. The rear side glass and tailgate are visually disjointed along the waistline leading from the main seating area; this implies there is a separate area for luggage which offers more space than rival models do. Adequate ground clearance and dark-grey lower trim hint at off-road potential but without being too rugged.

Inside, there is an unfortunate lack of inspiration. Tedious grey and dark polished wood mix with cream leather to create an interior design lacking the sumptuousness and tailored feel consumers expect in a top-end SUV. The instruments and switchgear are pleasingly straightforward, but, overall, the interior smacks of under-investment.

Chevrolet Aveo

Design	Italdesign
Engine	1.6 in-line 4
Power	78 kW (105 bhp) @ 5800 rpm
Torque	145 Nm (107 lb. ft.) @ 3600 rpm
Gearbox	5-speed manual
Installation	Front-engined/front-wheel drive
Front suspension	MacPherson strut
Rear suspension	Torsion beam
Brakes front/rear	Discs/drums
Front tyres	P185/60R14
Rear tyres	P185/60R14
Length	4235 mm (166.7 in.)
Width	1670 mm (65.7 in.)
Height	1495 mm (58.9 in.)
Wheelbase	2480 mm (97.6 in.)
Kerb weight	1080 kg (2381 lb.)

Designed at Giorgetto Giugiaro's Italdesign studio in Turin, the new Aveo from Chevrolet comes as either a contemporary four-door saloon or a sporty five-door.

The five-door body style has a bonnet that rises steeply on to the arched roofline. A strong side feature starts at the headlights and grows into a wedge that runs through to the rear lights. Both the 'drooping' feature in the lower half of the doors and the flared rear wheel arch add a little more visual interest to the side profile. Although the saloon has similar features, its door detailing is more horizontal, to reflect its greater length (the five-door model is 3800 mm/149.6 in. long).

Both cars are designed for mass-market appeal. They are totally inoffensive, mixing thoughtfully crafted proportions and attractive features to create models that are desirable for this marketplace but not too futuristic for the Chevrolet stable's sensible image. Points of interest include clear headlights with faceted lenses, jewel-like tail lights and prominent gold Chevy 'bowtie' badges front and rear.

Inside, the high roof and raised seats give good visibility and easy access, and are particularly welcome for their contribution to improving passive safety.

Chevrolet's marketers hope that launching the Aveo in the entry-level segment will allow the car to play an important role in luring new customers to the Chevy family. When it arrives at American dealers some time in 2004, the Aveo will be positioned to compete with the bargain-basement Hyundai Accent and Kia Rio, and it, too, is to be built in Korea to keep costs down and ensure ultra-competitive pricing.

Chevrolet Cheyenne

Design	Jeff Angeleri
Engine	6.0 V8 supercharged
Power	373 kW (500 bhp)
Torque	787 Nm (580 lb. ft.)
Gearbox	4-speed automatic
Installation	Front-engined/rear-wheel drive
Front suspension	Short and long arm
Rear suspension	IRS with rear steer
Brakes front/rear	Discs/discs
Front tyres	285/60R22
Rear tyres	285/60R22
Length	5940 mm (233.9 in.)
Width	2080 mm (81.9 in.)
Height	1945 mm (76.6 in.)
Wheelbase	3935 mm (154.9 in.)
Track front/rear	1770/1770 mm (69.7/69.7 in.)
Kerb weight	2721 kg (1234 lb.)

Chevrolet, which has been at the centre of the pickup truck's evolution for decades, has long prided itself on its rugged image. But the new Cheyenne concept may be a sign of things to come: it has a softer, more elegant style, plus a greater cargo-carrying versatility thanks to its huge load bed.

Certain Chevy trucks of the past, most notably the 1955 Cameo, the 1967 and 1973 Cheyennes and the 1988 Silverado, have made a great impression on General Motors' designers with their mix of design and function. The Cheyenne aims to capture the same iconic status.

Its cab is pushed further forwards, allowing more interior space, and thus greater versatility and comfort, and the wheels are pushed further outwards, giving the impression of increased stability. The two-panel glass roof with integrated sunroof and the wrap-around 'bubble back' rear glass accentuate the cab's openness.

The large cargo area could be quite tricky to reach, so Chevrolet has provided two side-access doors just behind the cab, as well as a traditional tailgate. There are multiple storage bins in the box floor, with lighting and integrated tie-downs for heavy jobs in all conditions.

Inside, the cab is simply surfaced and gives a sense of space. The instrument panel is low and uncluttered. Mottled saddle leather is used for the seats and headlining, while brushed satin-finish aluminium inlays in the doors and floor add a dash of quality.

This softened pickup attempts to inject some friendliness into a market that is crowded with tough, brutish-looking products. In doing so it may well find some new friends.

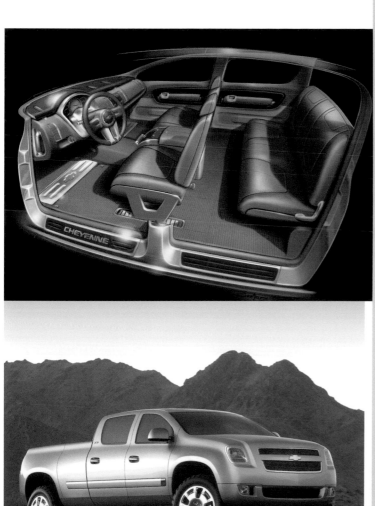

Chevrolet Colorado

Design	Clay Dean
Engine	3.5 in-line 5 (2.8 in-line 4 also offered)
Power	164 kW (220 bhp) @ 5600 rpm
Torque	305 Nm (225 lb. ft.) @ 2800 rpm
Gearbox	5-speed manual
Installation	Front-engined/rear-wheel drive
Front suspension	Independent with torsion bar
Rear suspension	Live axle with leaf springs
Brakes front/rear	Discs/drums
Front tyres	205/75R15
Rear tyres	205/75R15
Length	4887 mm (192.4 in.)
Width	1720 mm (67.7 in.)
Height	1648 mm (64.9 in.)
Wheelbase	2827 mm (111.3 in.)
Kerb weight	1541 kg (3397 lb.)

Chevrolet has an aggressive model programme launching new trucks, and the new Colorado marks the fifth all-new Chevy truck since the announcement of the Avalanche at the 1999 North American International Auto Show.

At the front the mix of lights, grilles, air intakes and body-coloured strips gives the Colorado an easily identifiable look, albeit one that is rather cluttered and resoundingly American rather than contemporary and global. An unapologetic gold Chevy 'bowtie' badge adorns the grille bar, reflecting the company's truck heritage. And chunky wheel arches with lots of wheel clearance clearly categorize this model as one that will be most at home off-road.

It must be a tough task convincing America's buying public that the GM stable has an excellent new engine that doesn't come in V8 form. The new 'I6' and 'I5' in-line five-cylinder units launched in the Chevrolet Trailblazer have made significant progress down this path, and it is derivatives of these engines that are used in the new Colorado. The earlier engines have passed on their particularly good power and torque curves – qualities that are ideal for this type of vehicle.

The Colorado comes in three body styles, all based on the same platform. In addition to the crew cab, the body style that accounts for one in three of all mid-sized pickups sold in the USA, there is the extended cab and the regular cab. The Colorado goes on sale during the last quarter of 2003 as an utterly traditional mid-sized pickup.

Chevrolet Equinox

Design	Ken Parkinson, Rich Scheer and Christos Roustemis
Engine	3.4 V6
Power	138 kW (185 bhp) @ 5200 rpm
Torque	285 Nm (210 lb. ft.) @ 3800 rpm
Gearbox	5-speed automatic
Installation	Front-engined/all-wheel drive
Front suspension	MacPherson strut
Rear suspension	Trailing arm
Brakes front/rear	Discs/drums
Front tyres	P235/65R16
Rear tyres	P235/65R16
Length	4757 mm (187.3 in.)
Width	1834 mm (72.2 in.)
Height	1681 mm (66.2 in.)
Wheelbase	2857 mm (112.5 in.)
Kerb weight	1634 kg (3602 lb.)

The new Equinox won't be seen until 2004, but the production version was unveiled early, at the Detroit show in January 2003. It comes complete with both the new Chevrolet look and unmistakable truck DNA.

The all-new Equinox has an altogether softer design than its slightly larger brother, the Colorado pickup, but the proportion is still classic SUV, with chunky features to match. A wide grille, big headlights and large pillars all make it look strong – and undoubtedly it is. The forward-angled C- and D-pillars give it a dynamic poise which, with a 3.4-litre V6 engine under the bonnet, it can rightly claim.

When launched, the Equinox will be the largest vehicle in its class, with a wheelbase of 2857 mm (112.5 in.). This gives it the advantage of better passenger comfort and cargo capability. One of the most useful features is Multi-Flex seating, where the rear seat slides by up to 20 cm (8 in.), providing extra room for passengers when moved back, or additional luggage capacity when moved forwards. The combination of a fold-flat front passenger seat with a split-folding rear seat also allows extra-long objects to be carried inside. Another useful feature is a height-adjustable cargo shelf that doubles as a security cover and a picnic table.

The Equinox will, no question, help Chevrolet put a strong new competitor into the fastest-growing market segment, compact utilities. This segment has breached the one-million sales point for the first time, and this is where the company needs to be.

Chevrolet Malibu

Design	Crystal Wyndham and Sung Paik
Engine	3.5 V6
Power	149 kW (200 bhp) @ 5400 rpm
Torque	285 Nm (210 lb. ft.) @ 3600 rpm
Gearbox	4-speed automatic
Installation	Front-engined/front-wheel drive
Front suspension	Strut type with stabilizer bar
Rear suspension	Independent 4-link with stabilizer bar
Brakes front/rear	Discs/drums
Front tyres	P215/60R16
Rear tyres	P215/60R16
Length	4783 mm (188.3 in.)
Width	1775 mm (69.9 in.)
Height	1461 mm (57.5 in.)
Wheelbase	2700 mm (106.3 in.)
Kerb weight	1504 kg (4820 lb.)

The Malibu is an important new model because it's the first US car based on General Motors' new Epsilon platform. In addition it is Chevrolet's new mid-sized saloon aimed at the mass market, and recalls a model name first introduced on a sportier iteration of the Chevelle, a 1960s best-seller for Chevrolet that occupied a similar market slot.

Its design is very unthreatening and conservative, to the point of being almost featureless. Large, taut surfaces end with squared-off corners and are intersected by simple shut-lines to make the overall form look boxy yet, with the chrome front bar complete with gold 'bowtie' badge, still distinctively Chevrolet.

Chevrolet promises that the Malibu will, due to its new platform, have excellent ride and handling. This unit is already used as the platform for the new Opel Vectra and Saab 9-3 models in Europe. The platform's design allows the ride to be softened slightly to suit America's more unpredictable road surfaces, so the chances are that the Malibu's road manners will not be as sharp as those of its European siblings.

There is an attempt to tailor the Malibu to the masses by offering greater comfort. This feature memorizes the seat height, steering wheel tilt/telescopic adjustment and the positions of the brake and accelerator pedals so that the driver's seating position can be programmed.

A first for this segment is the remote-start package, which allows the driver to start the car from inside the home. This could be a particularly useful feature on winter mornings or sweltering summer afternoons, and the system works over a distance of 60 metres (200 ft.).

Motorists who want to blend into the background should note that production of the Malibu is targeted for the third quarter of 2003. This is car design at its most underwhelming.

Chevrolet SS

Design	Franz von Holzhausen
Engine	6.0 V8
Power	320 kW (430 bhp)
Torque	584 Nm (430 lb. ft.)
Gearbox	4-speed automatic
Installation	Front-engined/rear-wheel drive
Front suspension	Short and long arm
Rear suspension	Short and long arm
Brakes front/rear	Discs/discs
Front tyres	255/45R21
Rear tyres	275/45R22
Length	5052 mm (198.9 in.)
Width	1930 mm (76 in.)
Height	1346 mm (53 in.)
Wheelbase	3150 mm (124 in.)
Track front/rear	1670/1651 mm (65.7/65 in.)
Kerb weight	1660 kg (3650 lb.)

The Chevrolet SS is one of those exotic and emotional concepts that most people drool over while going weak at the knees. Its four-door architecture has a rearward-biased cab design and is shrouded by clean-looking curvaceous bodywork. It's long, low and decidedly phallic.

With such a sports car, the wheels are very much the focus of attention. Sweeping wheel arches draw the hungry eye in, closely hugging the tyres and therefore signifying the limited suspension on offer and the hard, sporty ride that will surely result. At the back, circular tail lights project sportiness but in a modern way, with no fuss.

The SS, designed in GM's Los Angeles studio, is painted in victory red and fitted with reflective Cromax glass. The interior is similarly clean-looking, with hints of the heritage of the fabled Camaro SS, one of the first 'muscle cars' of the late 1960s. Stark off-white and slate leather surfaces contrast with the warm red exterior. The seat upholstery is also a blast from the past: a mixture of white leather and a modern woven, houndstooth-check vinyl.

A welcome technology for saving fuel is the SS's 'Displacement on Demand' system, which makes its debut in 2004 on some production models. This shuts down half of the cylinders during most driving conditions and automatically reactivates them for demanding use such as brisk acceleration or towing.

Chevrolet has a great sporting heritage with its iconic Corvette and Camaro. The SS, however, somehow looks more sophisticated than many of its past offerings. This is the closest thing yet to an American version of Ferrari's 456.

Chrysler Airflite

Design	Greg Howell and Simeon Kim
Engine	3.5 V6
Installation	Front-engined/rear-wheel drive
Front tyres	235/45R20
Rear tyres	255/45R21
Length	4838 mm (190.4 in.)
Width	1870 mm (73.6 in.)
Height	1448 mm (57 in.)
Wheelbase	2946 mm (116 in.)
Track front/rear	1600/1630 mm (63/64.2 in.)

Now that other Chrysler models, such as the Pacifica and the Crossfire, are in the hands of appreciative customers, the company's new design language is developing one stage further with the new Airflite concept. This sporty five-door coupé represents a fresh interpretation of Chrysler's automotive face, with its bold grille and distinctive headlights giving a strong impression of precision engineering.

The A-pillars rise sharply from a bonnet that boasts lively feature lines, three either side of the distinctive spine that runs the length of the car. From the top of the windscreen the roofline tapers down to the boat-tail at the back of the car, largely repeating the Crossfire's form. The tail is finished with a bold wing badge made of chrome.

In creating the Airflite, the designers took inspiration from contemporary furniture, classic boats and the Art Deco Chrysler Building in New York. Continuous radii and ovals are noticeably used in conjunction with straight lines for features and surfaces.

Nautical themes inspire the interior, but the overall look is chiselled and architectural – that is, functional but not particularly comfortable. A brushed-metal centre spine dominates, connecting the interior from front to back, but it contrasts sharply with the natural look of the wooden floor, which is accented with brushed aluminium strips to both protect the wood and echo the centre console. The satin silver-coloured centre console seems to form a structural member tying together the two sides of the car.

All a touch radical for now, perhaps, but Chrysler's concept cars have a pleasing habit of hitting the road if it's thought customers are willing to buy them.

Chrysler California Cruiser

You could be forgiven for thinking this is simply a PT Cruiser 'refresh', but in fact the California Cruiser is an all-new concept. It draws heavily, of course, on the design of the PT Cruiser, which has been on sale for three years, but it has a much more modern feel compared with its stablemate's deeply retro design.

The proportions are pretty similar to the PT's, but the California Cruiser is harder-edged, sporting contemporary silver body-side and tailgate panels. The rear is wrapped by several horizontal silver lines that give a crisp look and echo the prominent slats on the grille. One key feature of the concept, quite apart from its large load-carrying capacity, is the retractable hard-top. This has eight glass panels that lower or pivot to create fresh-air fun, as well as drop-down door glass and swivelling quarter-lights – a detail much loved by drivers with memories of 1950s cars – to let that LA sunshine flood in.

The California Cruiser shows a new face to Chrysler design, with scalloped headlights, a chrome-accentuated grille and an integrated bumper that reflects the detail design of both the new Crossfire and the soon-to-emerge Pacifica.

Inside, keynote horizontal elements are continued in the form of silver inserts on the door trim, quarter panels and hatchback lining, and this emphasizes the interior volume. For an ostensibly old-fashioned-looking car, the California Cruiser turns out to be pleasingly contemporary.

The modular seats have retractable head restraints, making the seats compact enough to be folded totally flat. Chrysler, however, never misses a trick for eye-catching trinkets: the winged Chrysler badge, in chrome, is set into the back of each seat.

The success of the PT Cruiser has been so important to an often-embattled Chrysler that the company is anxious to capitalize on any similar ideas; the California Cruiser proves it still has plenty of inspiration to mine.

Design	Trevor Creed
Engine	2.4 in-line 4
Gearbox	Autostick
Installation	Front-engined/rear-wheel drive
Brakes front/rear	Discs/discs
Front tyres	225/35R19
Rear tyres	225/35R19
Length	4290 mm (168.9 in.)
Width	1725 mm (67.9 in.)
Height	1511 mm (59.5 in.)
Wheelbase	2616 mm (103 in.)
Track front/rear	1481/1478 mm (58.3/58.2 in.)
Kerb weight	1454 kg (3206 lb.)

Chrysler Pacifica

Design	Trevor Creed
Engine	3.5 V6
Gearbox	4-speed automatic
Installation	Front-engined/all-wheel drive
Front suspension	MacPherson strut
Rear suspension	Multi-link
Brakes front/rear	Discs/discs, TCS
Front tyres	235/65R17
Rear tyres	235/65R17
Length	5052 mm (198.9 in.)
Width	2014 mm (79.3 in.)
Height	1689 mm (66.5 in.)
Wheelbase	2954 mm (116.3 in.)
Track front/rear	1676/1676 mm (66/66 in.)

In keeping with its growing tradition, Chrysler has taken another exciting concept vehicle and put it into production. Launched in buyer-ready form at the New York International Auto Show in 2002, the Pacifica represents a new vehicle segment: 'sports tourer'. Thanks to the design leadership of Trevor Creed, Chrysler now has an exciting range of cars in its showrooms and has tripled its market share over the past ten years. The Pacifica is the latest model in a recent line of innovative new products that includes the Prowler, PT Cruiser and Crossfire, and looks set to emulate their ground-breaking progress.

The production version of the Pacifica remains largely true to the original concept featured in *The Car Design Yearbook 1*, with only a few minor changes made to the headlight detail, door mouldings and handles. The lower body is now black, giving a marginally more rugged look despite the fact that the wheels are smaller. Chrysler believes there is nothing quite like the Pacifica in the marketplace, and while the BMW X5 offers strong competition, the Pacifica, with its low step-in height and three rows of seats, is a far more practical proposition for many Americans.

And this is a top-end car too. Luxury seating, a navigation system, hands-free communication and a power tailgate are all features to ensure the Pacifica can snatch sales from the best of the SUVs.

Citroën C-Airdream

Engine	3.0 V6
Power	157 kW (210 bhp) @ 6100 rpm
Gearbox	Auto-adaptive automatic
Installation	Front-engined/front-wheel drive
Front suspension	Hydractive
Rear suspension	Hydractive
Length	4499 mm (177.1 in.)
Width	1915 mm (75.4 in.)
Height	1289 mm (50.7 in.)
Wheelbase	2725 mm (107.3 in.)
Track front/rear	1747/1727 mm (68.8/68 in.)

Citroën's gorgeous C-Airdream concept is a beautifully crafted masterpiece. This design for a new sports coupé has a sleek, aerodynamic shape with a rearward-biased proportion and detailing that exudes sportiness and promises real enjoyment. You can't help thinking the spirit of the idiosyncratic but stunning 1970 Citroën SM has been crafted into a loving homage.

The car's 'face' is dominated by two large, slotted air intakes and narrow, swept-back headlights that sit either side of the famous double-chevron grille. The sculpted bonnet rises gently to the steeply raked windscreen, and this line continues up to the panoramic glass roof, giving occupants a feeling of complete openness. The roof panel gradually falls to the rear, ending abruptly as the body waists inwards to reduce aerodynamic drag – an SM touch for certain.

A subtle mix of dark, glossy and matt leather flows throughout the cabin, so that the interior oozes a sculptural surface language reminiscent of a sultry cocktail bar. The moiré-look seats are refreshingly different and look amply comfortable.

As with Citroën's previous concept car, the C-Crosser, there is a technology focus on drive-by-wire systems. On the C-Airdream, controls for accelerator, brakes and gear change are all located on the steering wheel, completely eliminating the need for a pedal box. This in turn gives more scope for interior design and improves safety. Even the steering actuation is via electronic control systems, obviating the need for conventional mechanical parts.

If only Citroën had the nerve to produce this car – as it did the SM – it would surely connect with a very chic audience, and bring to market a car that is in many ways truly new and exciting.

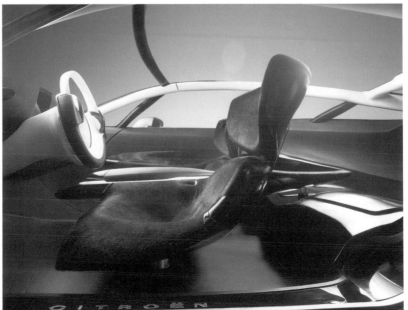

Dodge Durango

Design	Trevor Creed
Engine	5.7 V8
Power	257 kW (345 bhp) @ 5400 rpm
Torque	509 Nm (375 lb. ft.) @ 4200 rpm
Gearbox	5-speed automatic
Installation	Front-engined/four-wheel drive
Front suspension	Multi-link
Rear suspension	Solid axle, coil springs
Front tyres	265/50R21
Rear tyres	265/50R21
Length	5009 mm (197.2 in.)
Width	1943 mm (76.5 in.)
Height	1824 mm (71.8 in.)
Wheelbase	3028 mm (119.2 in.)
Track front/rear	1636/1636 mm (64.4/64.4 in.)

The Dodge Durango concept provides a strong hint of the next-generation Durango, which will appear in late 2003. Dodge says it represents 85% of the finished production-level exterior design. The current Durango is already a huge truck, but this new concept is at least 7.5 cm (3 in.) taller still, as well as being wider. It has more interior space and can carry seven people. The enlarged cargo deck can now hold standard-sized plasterboard, which is 122-cm (48 in.) wide – good news for builders.

Presence and power are easy to portray when you have a 5.7-litre engine and a truck of just over 5 metres (16.4 ft.) long. Even so, the massive cross-hair grille and snorting air intakes on the bonnet make this one of the most masculine-looking trucks on the market.

Dodge, however, has been diligent in ensuring the Durango is also sporty, and not too agricultural. Short overhangs, aluminium sill guards and roof rails and a raked windscreen all contribute, as does a gently rising waistline. The lower body highlights its off-road capability: protective guards front and rear, and distinctive, bulging wheel arches with plenty of tyre clearance for off-road suspension articulation.

The interior follows some clear geometric rules. Surfaces are constructed using constant radii for a simple overall form and, perhaps, a little rural ruggedness. But dark slate leather and an absence of wooden trim give the cabin a rather cold look.

Dodge Kahuna

The Kahuna concept is specifically designed for California's coastal culture. After all, Californians buy a lot of cars and are a fashion-conscious bunch. The Kahuna has enough space to seat six but can easily be adapted to carry more beach gear and fewer people.

There are some obvious design similarities to the Chrysler California Cruiser (see p. 64), but the Dodge is essentially a cool-looking people carrier, something that is often difficult to achieve because of the limitations of the necessarily boxy shape. Rather than a sliding rear door it has a conventionally hinged one, but with the upper frames and structure removed. This means the Kahuna can be driven totally open at the side when all the windows are dropped.

The Kahuna is a mono-volume design, to maximize interior space. The exterior is finished in Point Break blue, with accents of composite bird's-eye-maple laminate along the side panels. The sliding roof is made from a silver-grey canvas that is both water-resistant and transparent.

The front end features a pared-down interpretation of Dodge's cross-hair grille. Although just about recognizable, this design just doesn't have the beauty and finesse of the car's side profile, with its delicate horizontal lines. The chunky wheel arches look great, however, and project a youthful, go-anywhere spirit. Inside, the 'big wave' theme continues throughout, in two tones of blue. A wave design is used to shape the instrument panel as well as the switches, door-panel detailing and seats.

It would be fantastic to see fun design like this make it to car showrooms across the USA, but it's highly unlikely a business case could ever be made for such a limited-market car. A darned shame.

Design	Trevor Creed
Engine	2.4 in-line 4 turbocharged
Power	160 kW (215 bhp)
Gearbox	4-speed automatic
Installation	Front-engined/front-wheel drive
Front suspension	MacPherson strut
Rear suspension	Coil springs
Front tyres	255/50R22
Rear tyres	255/50R22
Length	4714 mm (185.6 in.)
Width	1976 mm (77.8 in.)
Height	1702 mm (67 in.)
Wheelbase	3099 mm (122 in.)
Track front/rear	1659/1659 mm (65.3/65.3 in.)

Dodge Magnum SRT-8

The Magnum SRT-8 Concept is a fitting addition to the design vocabulary of the company that gave us the iconic Ram truck and Viper sports car. This new sports tourer clearly reflects Dodge's American 'muscle' design background and comes complete with a Hemi engine.

Its on-the-road presence and projection of power are highlighted by the Dodge 'face', with its huge grille, and the low bodywork, which makes the doors, in particular, appear deep and strong. Overall, there is a chiselled form to the body, with large surfaces broken by two-dimensional lines. The result is a purposeful look created by changes in light reflection.

The wheels are a very strong feature: deeply dished chunky five-spokes lurking under distinctive wheel arches that embrace them. A wide-opening tailgate that hinges forward of the rear pillar offers proper cargo flexibility.

The seating position is 6 cm (2.4 in.) higher than in the conventional Dodge car range because people are now accustomed to SUV-style seating. Four competition-style instrument gauges with machined aluminium trim rings are tunnelled in, like the Dodge Viper's. The Magnum SRT-8 also features Viper-inspired racing-style pedals of drilled aluminium.

Interior leather is dark slate grey with ochre accents. Three brushed-aluminium spokes radiate from an aluminium ram's head in the centre of the steering wheel, with the vertical spoke replicating the split spokes of the exterior wheels.

With a vast array of platforms at Dodge's disposal, and such a well-refined design as this to attach to one of them, it's likely this model will see production.

Design	Trevor Creed
Engine	5.7 V8
Power	320 kW (430 bhp) @ 5600 rpm
Torque	651 Nm (480 lb. ft.) @ 4000 rpm
Gearbox	5-speed automatic
Front suspension	Short and long arm
Rear suspension	Multi-link
Front tyres	275/40R20
Rear tyres	295/40R20
Length	5021 mm (197.7 in.)
Width	1880 mm (74 in.)
Height	1505 mm (59.3 in.)
Wheelbase	3048 mm (120 in.)
Track front/rear	1600/1600 mm (63/63 in.)
Kerb weight	1808 kg (3986 lb.)

Ferrari Enzo

Design	Pininfarina
Engine	6.0 V12
Power	485 kW (660 bhp) @ 7800 rpm
Torque	657 Nm (484 lb. ft.) @ 5500 rpm
Installation	Mid-engined
Brakes front/rear	Discs/discs
Front tyres	245/40ZR19
Rear tyres	345/35ZR19
Length	4720 mm (185.8 in.)
Width	2035 mm (80.1 in.)
Height	1147 mm (45.2 in.)
Wheelbase	2650 mm (104.3 in.)
Track front/rear	1660 mm/1650 (65.4/65 in.)
Kerb weight	1365 kg (3009 lb.)
0–100 km/h (62 mph)	3.65 sec
Top speed	350 km/h (218mph)

Named after Ferrari's legendary founder (although industry rumours initially suggested that it was to be called the F60), the new Enzo is the latest V12 supercar from the Italian marque. Ferrari and Michael Schumacher have so dominated Formula One over the past four years that in 2002 their success led to a reassessment of the rules with the aim of making the sport more competitive. So there can hardly have been a better time for a dominant Ferrari to launch a new F1-inspired road car.

Pininfarina is the design consultancy responsible for every production Ferrari over the past four decades, with the exception of the 1973–80 Bertone-designed Dino 308GT4. Naturally, this great studio was commissioned to style this latest model.

As with all supercars of this nature, aerodynamics are intrinsic to the car's function and style; indeed, the Enzo is an extreme example of this emphasis. Its body is made from carbon-fibre and Nomex composite, allowing Pininfarina's designers the freedom to create optimized surfaces that would be absolutely impossible with conventional metal pressings.

At the front, the Enzo sports an F1-style raised nose and two massive intakes that channel air up to the brakes and over the windscreen. The cockpit is small, to minimize drag, and tapers at the rear on to a flat section before finishing abruptly at the rear edge. The absence of a conventional rear aerofoil is possible owing to the huge venturis sitting between the rear wheels. These speed up the escaping air and pull the car down hard to the road surface.

Inside, it is a tight squeeze, with acres of visible carbon-fibre that concentrates the mind on functional dedication rather than mere comfort.

This is an extreme Ferrari, dedicated to performance, and it does not have the beautiful aesthetics of models like the 360 Modena. A limited production run of 399 Enzos is planned, but if you want one it's already too late: they have all been sold.

Fiat Marrakech

Based on the Fiat Gingo, the open-top Marrakech would be brilliant fun in countries where it doesn't rain much – Morocco, for example, home to the city that inspired the car's name.

It was created by the Idea Institute in Italy, and is designed purely to get people excited about the possibilities for the new Gingo – a well-worn marketing stunt often revived when Fiat launches a major new car. At the debut of the Cinquecento in 1992, for example, a flotilla of eight 'fun' concept cars from Italian design consultancies was used to inject a little verve into the proceedings.

Design-wise, the Marrakech's exterior shares the bonnet, front grille and lights with the Gingo, but there is a new front bumper with what appears to be a large plastic protector in the centre to give the car a more adventurous feel. Along the sides there are no doors, just a high sill complete with a grab rail to provide stiffness to the body, to give protection in side impacts and to act as a useful support when climbing into the car.

The lower surfaces around the rear are almost identical to the Gingo's, but the lights have been modified to include circular treatments to the white reflectors. There's also a protective guard in the rear bumper to increase durability. The dashboard is taken straight from the Gingo but features new colours to key in with the bright-green exterior paintwork.

It would be marvellous to see fun cars like this on the streets, but the potential market is much too small for Fiat to consider making one. Still, one of the smaller coachbuilders might find the Marrakech feasible as a low-volume, high-price successor to the long-lamented Mini Moke and Citroën Mehari.

| Design | Idea Institute |

Ford 427

Engine	7.0 V10
Power	440 kW (590 bhp) @ 6500 rpm
Torque	690 Nm (509 lb. ft.) @ 5500 rpm
Gearbox	6-speed manual
Front tyres	245/45R19
Rear tyres	245/45R19
Length	4760 mm (187.4 in.)
Width	1846 mm (72.7 in.)
Height	1344 mm (52.9 in.)
Wheelbase	2872 mm (113 in.)

Cool American cars such as the Ford Galaxie and LTD enjoyed their heyday in the 1960s. These days hatchbacks and SUVs dominate the scene. Yet, with the 427, Ford has chosen to remind us that it's still possible to create a modern interpretation of the desirable 1960s saloon by using classic proportions: long and low as you can get, with a distinctive three-box shape, and with large, uninterrupted surfaces separated by gentle lines.

Black is excellent for presenting a car that has such a terrifying presence, although the silver trim adds a touch of humanity. The wide grille emphasizes the width of the car and runs outboard to the vertical rectangular headlights. The bonnet lines are taut and hostile, and everything is magnified by the squareness of the proportions.

Inside, there's no let-up to the sinister nature of this car. Black leather with white stitching and brushed aluminium and chrome mix to highly dramatic effect. Gently flowing surfaces and a lack of clutter harmonize well with the exterior design.

The instrument cluster has a TV-screen-shaped speedometer and tachometer with a design that relates directly to that of the front and rear lights. These analogue gauges look frustratingly unclear and their red glow is prioritized well ahead of any functionality.

It's bewildering how, in a country where the speed limit is mostly no more than 97 km/h (60 mph), a 7-litre car can be anything but ridiculous, but with retro design so recently fashionable, the 427 definitely got its share of attention at the Detroit show. Which is, of course, its raison d'être.

Ford F-150

Design	Craig Metros
Engine	5.4 V8
Power	224 kW (300 bhp)
Installation	Front-engined/all-wheel drive
Front tyres	P215/60R16
Rear tyres	P215/60R16

The new F-150 from Ford is the replacement for a model that has led the sales leagues in full-sized pickups for an amazing twenty-six years. The F-150 has long been the USA's best-selling passenger vehicle, and Ford wants the new one to be regarded as not just a tough, utilitarian workhorse but also a comfortable and refined automobile.

The exterior is chunky and powerful, with bold proportions and features. At the front the large, inverted trapezoidal grille and square headlights dominate, together with a blue 23-cm (9-in.) oval badge, which looks slightly oversized at first but is in fact designed to be in proportion to the huge front end.

Deriving some inspiration from the 2002 F-350 Tonka concept truck, the F-150 uses similarly taut and chiselled shapes. But it features a 'waterline': a line extending the length of the vehicle and joining the tops of the front and rear bumpers to create a visual separation around the entire vehicle. This waterline creates visual differentiation between the F-150's five different production ranges – for example, as part of a two-tone paint treatment – while maintaining a design coherence.

The tailgate uses torsion-beam technology, so when it's lowered it feels more like lightweight aluminium or a composite panel than the relentlessly old-tech steel-reinforced tailgate it actually is.

Inside, there are vertical bands, which again allowed the design team to use different colours, textures and materials for range customization. The theme is quite contemporary and precise, with masses of aluminium and black, while the instrument design and layout faintly recall an aircraft cockpit.

Ford Model U

It's doubtful that a quantum leap in popular cars as large as that made by the Ford Model T in 1908 could ever occur again, but Ford thinks the Model U is just as visionary. This is a design exploration for a car that is likely to hit the road a hundred years after the immortal 'Tin Lizzie' made its debut. Whether or not you concur with such a grand design gesture, the Model U is certainly one of the most innovative concepts launched in 2003.

The Model U concept doesn't just explore user-functionality; it also considers its own effect on global resources by looking at modular design, recyclable materials, emissions and disassembly.

The exterior three-box profile is tough yet funky, much more successful than the Freestyle concept, but perhaps not the right product just yet for a Ford customer, who may not be ready for anything quite so radical.

The body panels have different finishes and are made of different recyclable materials. The body side is glossy, yet the doors are matt, with grooves to make them appear more structural. The body structure is aluminium and the front side panels are made of a natural fibre-filled composite.

There are many innovations, but the supercharged hydrogen internal-combustion engine is a real world first, although it has not yet been independently validated. The interior is modular, so it can be upgraded as the owner chooses. Also inside, a conversational speech interface allows a person to speak to operate on-board systems, including entertainment, navigation, cellular telephone and climate control. Pre-crash sensing, adaptive front headlights and a night-vision system help the driver to avoid accidents.

Advanced materials are designed for their ecological effects and can go from cradle to cradle, instead of staying in the cradle-to-grave waste streams typical of the car industry. Rubber tyres use corn-based fillers as a partial substitute for carbon black.

Such innovations are fundamental to the progression of design. In the absence of tougher legislation, more V8-powered trucks will find their way to the marketplace without really challenging designers and engineers to come up with solutions like those embodied in the Model U.

Design	Laurens van den Acker
Engine	2.3 in-line 4 hydrogen-powered
Power	113 kW (152 bhp) @ 4500 rpm
Torque	210 Nm (155 lb. ft.) @ 4000 rpm
Brakes	Discs/discs
Length	4230 mm (166.5 in.)
Width	1810 mm (71.3 in.)
Height	1651 mm (65 in.)
Wheelbase	2685 mm (105.7 in.)
Track front/rear	1583/1583 mm (62.3/62.3 in.)
Fuel consumption	6.3 ltr/100 km (45 mpg)
CO_2 emissions	PZEV

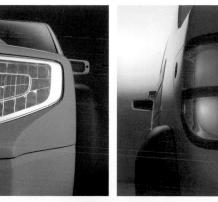

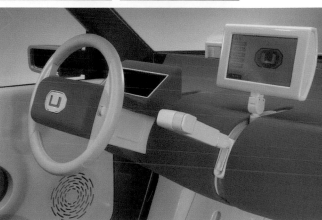

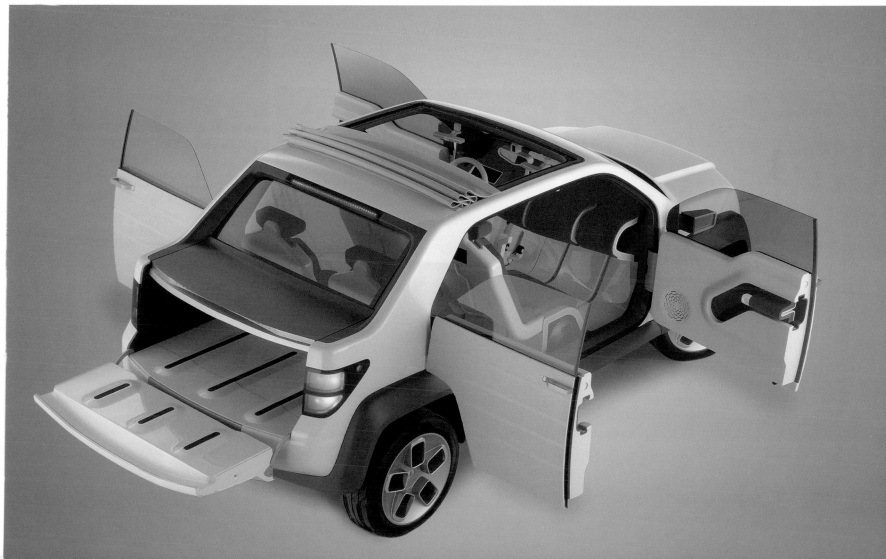

Ford Streetka

Design	David Wilkie
Engine	1.6 in-line 4
Power	70 kW (94 bhp) @ 5500 rpm
Torque	135 Nm (99 lb. ft.) @ 4250 rpm
Gearbox	5-speed manual
Installation	Front-engined/front-wheel drive
Front suspension	MacPherson strut
Rear suspension	Twist beam
Brakes front/rear	Discs/drums
Front tyres	195/45R16
Rear tyres	195/45R16
Length	3650 mm (143.7 in.)
Width	1695 mm (66.7 in.)
Height	1335 mm (52.6 in.)
Wheelbase	2448 mm (96.4 in.)
Track front/rear	1417/1452 mm (55.8/57.2 in.)
Kerb weight	1061 kg (2339 lb.)
0–100 km/h (62 mph)	12.1 sec
Top speed	173 km/h (108 mph)
Fuel consumption	7.9 ltr/100 km (35.8 mpg)
CO_2 emissions	190 g/km

The Streetka has been eagerly awaited since the concept for it was first unveiled at the Turin Motor Show in 2000. The original, racy-looking design – christened Saetta – was designed by Ghia, Ford's venerable design house in Turin, which since then has virtually closed down. It featured a radically sporty exterior and a cockpit bursting with bright aluminium and red fabric.

The production model has been slightly tamed. Designed for production and built by Pininfarina in Italy, the latest Streetka broadly retains the concept's exterior style. At the front the edgy design translates into sharp new headlights, while the rear has new lights that triangulate downwards, eventually meeting the single exhaust pipe in the centre. The result is a customer-ready model that is less overt yet still distinctly recognizable as a Ka, thanks to its blistered wheel arches and rounded profiles. The concept's interior, however – and alas – has been substantially toned down to a more conventional Ka one, no doubt to reduce project costs.

The Streetka is aimed at fashion-conscious, pop-music-addicted twenty- and thirty-somethings. With that in mind, Ford has recruited Kylie Minogue to promote the car, which gives you an idea of where the Streetka brand is heading.

Surprisingly, the Streetka is the smallest convertible yet produced by Ford, although it is not the first Ford to be built in Italy: in the early 1960s an Italian-assembled and restyled version of the Ford Anglia 105E called the Anglia Torino was built. If only Ford had predicted the Streetka's rapturous reception – it is probably kicking itself for not getting it to market faster, when the Ka itself was still fresh out of the box. Still, the rub-off will surely strengthen the Ka brand and boost sales of all Ka derivatives.

Honda Accord

Design	Honda Motor Corporation
Engine	3.0 V6 (2.4 in-line 4 also offered)
Power	179 kW (240 bhp) @ 6250 rpm
Torque	288 Nm (212 lb. ft.) @ 5000 rpm
Gearbox	5-speed automatic
Installation	Front-engined/front-wheel drive
Front suspension	Double wishbone
Rear suspension	Double wishbone
Brakes front/rear	Discs/discs, ABS, EBD
Front tyres	205/60R16
Rear tyres	205/60R16
Length	4813 mm (189.5 in.)
Width	1816 mm (71.5 in.)
Height	1450 mm (57.1 in.)
Wheelbase	2740 mm (107.9 in.)
Track front/rear	1552/1554mm (61.1/61.2 in.)
Kerb weight	1524 kg (3360 lb.)

It is a privileged job having to redesign the best-selling car in America, but the task must come with a good share of worries. How far to progress the design to demonstrate it is up to date while protecting loyal customers and market share? In the USA alone, more than eight million Accords have been sold since the model went on sale in 1976; that is some achievement, but a weighty legacy for designers.

Honda's design centre in Japan says that the difference between Japanese and European customer expectations has become much smaller, but the US market still requires larger engines and unique exterior styling. Therefore, cars made for Europe and Japan appear to be identical, but versions made for America are fractionally wider and longer, to accommodate a V6 engine. There is a coupé too, but it is offered only in the USA.

The new model is altogether crisper than its predecessor. The headlights now have sharp outlines and the nose a more chiselled look. The side profile has a high-rising waistline to project strength and sportiness, with an abruptly truncated rear similar to those produced by manufacturers like Alfa Romeo and Audi. Pull door handles and mirror-mounted indicators add to the perception of quality.

Hondas always score highly in the influential JD Power quality surveys in the USA, so, irrespective of how radical the new design was going to be, sales of Honda's new Accord were always certain to be enormous. And the company has managed to add a tiny bit of 'edge' to one of the most loyally bought of all cars.

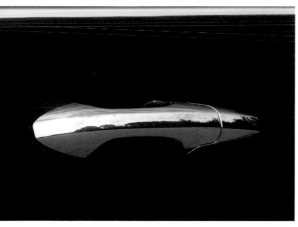

Infiniti M45

Engine	4.5 V8
Power	254 kW (340 bhp) @ 6400 rpm
Torque	452 Nm (333 lb. ft.) @ 4000 rpm
Gearbox	5-speed automatic
Installation	Front-engined/rear-wheel drive
Front suspension	MacPherson strut
Rear suspension	Multi-link
Brakes front/rear	Discs/discs TCS, BA, EBD
Front tyres	235/45R18
Rear tyres	235/45R18
Length	5009 mm (197.2 in.)
Width	1770 mm (69.7 in.)
Height	1463 mm (57.6 in.)
Wheelbase	2799 mm (110.2 in.)
Track front/rear	1509/1509 mm (59.4/59.4 in.)
Kerb weight	1747 kg (3851 lb.)
Fuel consumption	9.8 ltr/100 km (28.8 mpg)

The all-new, V8-powered Infiniti M45 is positioned between Nissan's flagship Q45 and the new Infiniti G35 Sport Sedan.

Infiniti is the manufacturer's premium brand, tilted at a primarily US customer base. Unlike Nissan, which now has a strong product character with the new Primera and Micra thanks to input from Renault's head of design, Patrick le Quément (Renault is Nissan's majority shareholder), Infiniti has stuck to its conservative and inoffensive guns. There is nothing wrong with the M45's understated exterior, but it suggests a comfortable cruising car rather than an outright performer with V8 firepower in its belly. It is cautious where a Mercedes-Benz S-Class exudes excitement.

The M45 has a more squared-off appearance than the smaller G35. Its front end is dominated by an aggressive front grille, with wide, black-chrome horizontal bars that flow into the moderately sporty xenon headlights. Viewed from the side, the silhouette is a wedge shape running front to rear, with a swept-back C-pillar and a steeply running off boot that contrasts with an upright A-pillar.

Inside, the contoured front seats are finished with 'Sojourner' leather exclusive to the Infiniti brand. The interior also features a bird's-eye-maple trim with a smoked-graphite colour, and a performance-orientated four-dial instrument cluster complete with amber lighting that silhouettes the gauges.

Progressive technology is strongly featured too. There is a vehicle information system consisting of a large, multi-function LCD screen and an optional DVD-based navigation system, an intelligent cruise control that maintains driver-selected following distances and the Infiniti voice-recognition system shared with the Q45.

The M45 is a dignified and technically accomplished luxury car, then, but a bit of an identity-free zone.

Invicta S1

Design	Leigh Adams
Engine	4.6 V8
Power	239 kW (320 bhp) @ 5900 rpm
Torque	407 Nm (300 lb. ft.) @ 4800 rpm
Gearbox	5-speed manual
Installation	Front-engined/rear-wheel drive
Front suspension	Double wishbone
Rear suspension	Double wishbone
Brakes front/rear	Discs/discs
Front tyres	255/35ZR19
Rear tyres	275/35ZR19
Length	4400 mm (173.2 in.)
Width	2000 mm (78.7 in.)
Height	1225 mm (48.2 in.)
Wheelbase	2500 mm (98.4 in.)
Track front/rear	1780/1730 mm (70.1/68.1 in.)
Kerb weight	1098 kg (2420 lb.)
0–100 km/h (62 mph)	5 sec
Top speed	274 km/h (170 mph)
Fuel consumption	11.3 ltr/100 km (25 mpg)

Invicta is a revival of a marque that has not been represented in the price lists since 1950, and whose heyday was a good fifteen years before that. Moreover, a resurrection of the name (which comes from the White Knight of Edmund Spenser's late sixteenth-century allegorical romance *The Faerie Queene*) has been attempted several times in the intervening period.

The first Invicta of modern times, therefore, should be an intriguing confection. And the new Invicta Car Company has harnessed the design ethos of the original pre-war designers like Reid Railton and William Watson, who ensured that every component projected a masculine message of speed, power and visual drama.

A high waistline and a low overall height combine to give a somewhat cocooned and secure feeling inside the S1. The interior is sumptuous, with wall-to-wall tan and grey leather, and metallic detailing on instruments, centre console and pedals.

Outside, the Invicta winged badge straddles the inset bonnet and 'smiling' grille, while the front apron drops at the edges to give more coverage to the front wheels. This feature of dropped skirting is mirrored at the sides on the sills and on the rear apron.

In the 1930s Invictas were robustly engineered, with a quality that closely matched Rolls-Royce, and engines that gave enormous torque outputs. The S1's construction, though, is unique: it is the world's first production car with a one-piece carbon-fibre bodyshell. This is bonded to a steel tubular spaceframe chassis to create a very strong, stiff and lightweight structure that gives high torsional rigidity for good roadholding and excellent occupant-cell integrity for enhanced safety. With power from Ford's V8 Mustang engine, the S1 promises to be rapid, safe and stylish – something Invicta's founder, Noel Macklin, would be proud of.

Of course, the S1 faces tough competition from prestige marques like Porsche and Maserati. But, hopefully, for the sake of automotive biodiversity, it will rekindle this vintage name while bringing exciting motoring to an exclusive coterie of owners.

Jaguar XJ

Design	Ian Callum
Engine	4.2 V8 (3.0 V6 and 3.5 V8 also offered)
Power	298 kW (400 bhp) @ 6100 rpm
Torque	553 Nm (408 lb. ft.) @ 3500 rpm
Gearbox	6-speed automatic
Installation	Front-engined/rear-wheel drive
Front suspension	Double wishbone
Rear suspension	Double wishbone
Brakes front/rear	Discs/discs
Front tyres	255/40R19
Rear tyres	255/40R19
Length	5080 mm (200 in.)
Width	1868 mm (73.5 in.)
Height	1448 mm (57 in.)
Wheelbase	3034 mm (119.4 in.)
Track front/rear	1556/1546 mm (61.3/60.9 in.)
Kerb weight	1665 kg (3671 lb.)
0–100 km/h (62 mph)	5.3 sec
Top speed	249 km/h (155 mph)
Fuel consumption	12.3 ltr/100 km (23 mpg)
CO_2 emissions	299 g/km

This new XJ from Jaguar is the seventh generation of its flagship saloon first introduced in 1968. While in many ways a subtle evolution of thirty-four years of timelessness, it is also the most radical rendition to date – on the inside.

It is longer, taller and wider than its predecessor, giving more interior room, although the iconic shape remains largely the same. Subtle changes include deeper doors, a higher waistline and a marginally more 'cab-forward' package; the front overhang is reduced and the windscreen rake is faster, but the smaller bonnet still retains the characteristically sculpted XJ shape, initiated from the oval headlights. The headlights are set either side of a new grille, with intersecting vertical and horizontal bars, which takes its inspiration from the original 1968 model.

As with all XJs in the past, the latest technology features strongly. But the most advanced new feature is a choice of material. The body is made from riveted and bonded aluminium, resulting in a body-weight reduction of around 40% and shaving a hefty 200 kg (441 lb.) off the complete car.

The interior styling uses gently curved surfaces and quality materials to give the kind of comfortable, executive atmosphere you would expect from a top-of-the-range Jaguar. The sporty versions have a combination of charcoal dashboard and grey-stained bird's-eye-maple veneer, while classic models have traditional burr-walnut finish. A new option is piano-black trim, a highly polished finish used on the centre console and gearshift surround.

Inside, another new technology enhances safety: JaguarVoice is the voice-activated control of the audio system, telephone, climate control and navigation.

Kia KCV-II

A funky new crossover concept from Kia, the KCV-II is intended to drum up some new excitement for this sometimes lacklustre brand by targeting youthful car owners who enjoy an active lifestyle. Kia says the target audience is the 'Bobo' generation, which stands for 'bourgeois bohemians'. These are, we are told, people who are economically bourgeois but spiritually bohemian – a mixture of 1960s hippies and 1980s yuppies.

The KCV-II has an all-new platform designed so that it can accommodate several body styles, from a conventional sports utility vehicle to a lifestyle variant like the KCV-II concept that Kia has come up with. With high ground clearance, a commanding driving position and an open pickup cargo space, the KCV-II has the stance and tough looks of an SUV.

The body's curvaceous surfaces make it look softer, in an effort to target female bobos. Non-automotive products have been used for inspiration throughout: design elements from motorcycles, furniture, architecture and consumer products are fused together here.

A striking feature is the aluminium band that circles the car, starting at the nose, running along the waist and finishing as a guard rail over the pickup load bed. This highlights the wedge of the profile – a classic styling feature used to imply dynamism in a car's look.

The KCV-II's 'scissor' doors would more usually be found on a Lamborghini – the Countach, Diablo or Murciélago, as well as Toyota's short-lived 1990s Sera – and, unsurprisingly, give a sporty feel to the body architecture. They also have the practical effect of improving access to the rear seats.

The interior has a warm mix of pastel leather shades and precision-machined surfaces. Instruments are clustered around the steering column, their design inspired by motorcycle instrument binnacles.

For those in the back, the rear glass screen hinges upwards so that when the rear seats are reclined the passengers are partly exposed to the elements, just as in a cabriolet.

Design	Jay Baek and Peter Arcardipane
Engine	3.5 V6
Torque	294 Nm (217 lb. ft.) @ 3000 rpm
Gearbox	5-speed automatic
Installation	Front-engined/4-wheel drive
Front suspension	MacPherson strut
Rear suspension	Strut with trailing arm
Brakes front/rear	Discs/discs
Front tyres	255/50R20
Rear tyres	255/50R20
Length	4535 mm (178.5 in.)
Width	1860 mm (73.2 in.)
Height	1820 mm (71.7 in.)
Wheelbase	2615 mm (103 in.)
Track front/rear	1590/1660 mm (62.6/65.4 in.)
Kerb weight	1620 kg (3571 lb.)
0–100 km/h (62 mph)	8.5 sec
Top speed	210 km/h (130 mph)

Lincoln Aviator

Design	Gerry McGovern
Engine	4.6 V8
Power	225 kW (302 bhp) @ 5750 rpm
Torque	407 Nm (300 lb. ft.) @ 3250 rpm
Gearbox	5-speed automatic
Installation	Front-engined/rear-wheel drive or all-wheel drive
Front suspension	Independent short and long arm
Rear suspension	Independent short and long arm
Brakes front/rear	Discs/discs, EBD
Front tyres	245/65R17
Rear tyres	245/65R17
Length	4910 mm (193.3 in.)
Width	1930 mm (76 in.)
Height	1814 mm (71.4 in.)
Wheelbase	2888 mm (113.7 in.)
Track front/rear	1547/1554 mm (60.9/61.2 in.)
Kerb weight	2180 kg (4805 lb.)
Fuel consumption	12.3 ltr/100 km (23 mpg)

The new Lincoln Aviator has come out of the same mould that produced the Navigator last year, although it is slightly smaller. Such is the consistency of the design, this new model is quite simply a mid-sized replica of its rather gross brother. With the Navigator, Lincoln already holds a leading 40% share of the premium SUV market, so it hopes that adding this new mainstream version will inflate its market share significantly.

The chromed signature waterfall grille, with its dark vanes, and large, clear-lens headlights dominate the Aviator's 'face'. The lower bumper adds to its aggressive stance, with body-coloured cladding wrapping around the wheel arches and lower sections of all four doors to protect against chips, dents and door dings. Full-length running boards are integrated into the sill but are not powered like the Navigator's. Chrome is carefully used to pick out the grille surround, waistline, roof rack and rear number-plate surround.

Inside, the design direction is also taken from the Navigator, with the same blend of Lincoln trade-marks – satin nickel, American walnut burr wood, leather and more than a hundred white light-emitting diodes that illuminate the instruments and most buttons and controls and complement the satin-nickel finish. The prominence of the analogue clock is now becoming a traditional touch across all Lincoln-brand cars. A band of wood works its way around the cabin, providing a visual waistline to draw the eye to its extremities and making it seem more spacious.

The 'Russian doll' approach has worked for other manufacturers, notably BMW with its 3, 5 and 7 Series family. And it should prove a lucrative strategy for Lincoln.

Mitsubishi Tarmac Spyder

Engine	2.0 in-line 4
Power	224 kW (300 bhp)
Gearbox	5-speed automatic
Installation	Front-engined/all-wheel drive
Front tyres	225/35R19
Rear tyres	225/35R19
Length	4055 mm (159.7 in.)
Width	1825 mm (71.9 in.)
Wheelbase	2515 mm (99 in.)

Youthful and extreme, the Tarmac Spyder concept is a two-plus-two aimed at the techno-savvy young 'Generation Y' market, and is a follow-up to the CZ-3 Tarmac sports hatchback shown in 2001. Supposed to be fun to drive, the car has racy lines; proportionally, it's a dramatic wedge that begins at the front grille and rises sharply up to the waistline and on to the boot lid.

At the front there is a stark contrast between the narrow 'arrowhead' headlights alongside the bonnet and the wide-open 'mouth' that is built into the front bumper. This mouth makes the car look aggressive and manly, but it's very necessary for the massive cooling needed by the engine. The door mirrors echo the car's overall spearhead shape.

The Spyder comes with a detachable hardtop, which lets Mitsubishi offer rear seats as standard. More acceptable today is to have a folding roof and a retractable hardtop, and make the car a dedicated two-seater. The Ford Streetka (see p. 110), although only offered as a traditional soft-top, would be a competitor if the Spyder were ever produced, because it's a great example of a dedicated two-seater package.

The dashboard and instrument panel use wave-line styling themes. A camera mounted in the door mirrors can record a journey or play it on individual LED monitors to the front and rear passengers. The interior scheme is racing-car-inspired, with body-hugging seats and a highly functional aura.

Nissan Quest

Engine	3.5 V6
Power	172 kW (230 bhp)
Torque	319 Nm (235 lb. ft.)
Gearbox	5-speed automatic
Installation	Front-engined/front-wheel drive
Front suspension	MacPherson strut
Rear suspension	Multi-link
Brakes front/rear	Discs/discs, ABS, BA, EBD
Front tyres	225/60R17
Rear tyres	225/60R17
Length	5184 mm (204.1 in.)
Width	1971 mm (77.6 in.)
Height	1778 mm (70 in.)
Wheelbase	3150 mm (124 in.)
Track front/rear	1699/1699 mm (66.9/66.9 in.)

Trying to create an object of desire when designing a minivan is one of the trickiest tasks in car design today. Compromise on functionality and your car will be massively criticized, because function is deemed to be a minivan's prime purpose. For this reason the results are often unprepossessing boxes on wheels that stimulate precious little emotion in onlookers or, crucially, potential owners. To put it bluntly, most MPVs are boring.

For the Quest, Nissan's designers have really tried to get round this problem, and have gone for a one-and-a-half box shape that stretches out to a length of 5.2 metres (17.9 ft.) but retains curvaceous design cues in its roofline and in the waistline, which is initiated by the headlights and then weaves its way backwards, visually separating the forward cabin from the rear passenger space.

At the front, the grille and headlights are unmistakably the new face of Nissan, while the rear lights sit outside the tailgate, emphasizing the car's width.

Although sliding rear doors must be upright enough to allow the door and its mechanism to slide outside the fixed body, the Quest's design manages to include some curved form in them and gives the wheel arches some flair, to project a certain sportiness.

The contemporary interior is noteworthy: the Skyview roof has four panoramic glass windows and a rear overhead aircraft-style console with personal reading lights and air vents, as well as twin DVD display screens. Other innovations include a low-height instrument panel, set in the centre of the dashboard, that also contains the gearshift and myriad storage compartments.

This is a sterling effort at putting some visual interest into an everyday car without chipping away at its practicality.

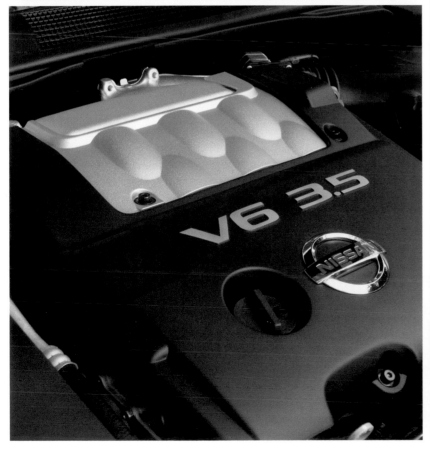

Nissan Titan

Engine	5.6 V8
Power	224 kW (300 bhp)
Torque	509 Nm (375 lb. ft.)
Gearbox	5-speed automatic
Installation	Front-engined/4-wheel drive
Front suspension	Double wishbone
Rear suspension	Leaf spring
Brakes front/rear	Discs/discs, ABS
Front tyres	265/70R18
Rear tyres	265/70R18
Length	5695 mm (224.2 in.)
Width	2002 mm (78.8 in.)
Height	1890 mm (74.4 in.)
Wheelbase	3550 mm (139.8 in.)
Track front/rear	1715/1715 mm (67.5/67.5 in.)

Titan is Nissan's first-ever model in the full-sized truck category, and will be offered late in 2003 as both King Cab and Crew Cab derivatives.

The Titan name was chosen to reflect the marketing importance Nissan stakes on the truck's size and power. The company's own research has concluded that the key elements in the successful design of a full-sized pickup continue to be horsepower, torque, towing capacity, a large cab and an enormous load bed.

Designed at Nissan's studio in California, the Titan has a distinctive front end with large proportions. One detail that differentiates it from other models is the expanse of chrome on the grille and bumper. The bonnet, which has a large, raised centre section designed to clear the engine, is quite flat in profile. The sides are uncluttered apart from chrome detailing on the door handles, the door-mirror caps and the proudly worn V8 badge.

A couple of very useful innovations should smoothe Nissan's debut in a previously untapped segment. The load bed comes complete with a factory-installed spray-in bed liner to protect it from scratches and eventual corrosion. Also, the Titan has no B-pillars, so when both doors are open there is easy access to all the interior space.

Inside, the cab is spacious and straightforward, with clearly laid-out controls. The door handles, knobs and assist grips are all designed specifically for the Titan, and can be used with work gloves on; it's a small detail, and more costly for Nissan than picking existing car-like components out of its enormous parts bin, but it will definitely please Titan users.

Opel GTC Genève

'GTC' stands for Gran Turismo Compact and is a fashionable enough abbreviation, Opel hopes, to make us believe that it has discovered an all-new vehicle segment. This concept car is, in fact, a three-door coupé and so isn't a novelty; in 1999 Opel's American cousin Saturn put a three-door coupé, the SC2, on sale in the USA. Even so, it's new for Europe, and Opel could claim that the GTC Genève's large-car underpinnings give it the ride qualities to justify its grand name.

The exterior design language uses taut lines, uninterrupted surfaces, wide wheel arches and strong features – all already seen in studies like the Snowtrekker and in production models like the Speedster and the Vectra.

With new dynamic design language, the GTC Genève appears to mix design elements from other marques, like the centre-line feature on the front bumper and bonnet, the small rear quarter-light and the strong C-pillar – all reminiscent of Renault. This car looks individual, but it's really following the design fashions of today rather than forging them.

A tinted transparent roof, stretching from the windscreen to the rear window, lies on top of a curved roofline that creates a powerful tension that is visible from the side. The clear headlights and tail lights are jewel-like, and mounted centrally above the rear window are two rather over-designed brake lights.

The interior is sporty, with upholstery of cashmere and brown leather. The cockpit has charcoal-coloured surfaces and some even darker ones, to give the lie to the modern cliché that aluminium best suggests sportiness.

Design	Martin Smith and Friedhelm Engler
Front tyres	245/40R19
Rear tyres	245/40R19
Length	4349 mm (171.2 in.)
Width	1773 mm (69.8 in.)
Height	1352 mm (53.2 in.)

Opel/Vauxhall Meriva

Design	Friedhelm Engler
Engine	1.8 in-line 4 (1.6 in-line 4 and 1.7 in-line 4 diesel also offered)
Power	92 kW (123 bhp)
Torque	165 Nm (122 lb. ft.) @ 4600 rpm
Gearbox	Easytronic
Installation	Front-engined/front-wheel drive
Front suspension	MacPherson strut
Rear suspension	Torsion-crank axle
Brakes front/rear	Discs/discs, ABS, BA
Front tyres	185/60R15
Rear tyres	185/60R15
Length	4040 mm (159.1 in.)
Width	1694 mm (66.7 in.)
Height	1620 mm (63.8 in.)
Wheelbase	2630 mm (167.3 in.)

In 2003 the new Meriva will not replace Opel's – or, in the UK, Vauxhall's – mainstay MPV, the Zafira, but will instead complement its big brother as an ultra-compact people-carrier. It relates to the Corsa in the same way that the Zafira shares some engineering with the Vauxhall Astra, and will be built alongside the Corsa at the Zaragoza plant, in Spain, for Europe. The other source will be San José dos Campos, in Brazil, for the South American market, where it will be called the Chevrolet Meriva; General Motors aways likes to extract as much from each new vehicle as possible.

A five-seater minivan, rather smaller than the Zafira, the Meriva is quite understated and inoffensive, despite being curvy with short body overhangs and curved roof pillars that continue over to the rear pillars. The overall proportions are of a cab-forward design with a curved back, like a scaled-down Citroën Picasso. A rising feature line runs from the headlights right to the rear, creating dynamic shoulders that portray strength.

The established Opel face is more prominent now, with a chrome band across the trapezoidal grille carrying the lightning-flash motif, while the tail lights sit neatly in the pillars alongside the rear screen.

Inside, the colour scheme is dark grey. The main design objective, however, was the back-seat concept, a folding system that turns the rear passenger area into a flat-floored cargo space that would not shame a small delivery van.

Available as an option are so-called 'infotainment' systems providing information, music and video. In the Meriva, this translates as in-car navigation, a car phone or the OnStar telematics service.

Opel believes that this low-cost minivan will, by 2005, have taken a considerable share of a half-million-unit segment. On this basis, GM is playing pretty safe with its design execution.

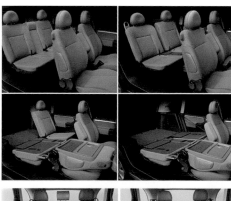

Peugeot H₂O

Sometimes an unconventional idea like the H₂O actually makes a lot of sense. At first sight this is simply a whimsical concept for a compact fire-fighting vehicle, despite Peugeot's assertion that the H₂O was designed in conjunction with French firemen so that it can, uniquely, fit into tight spaces inaccessible to conventional fire engines.

But look more closely and the true rationale for the H₂O concept becomes apparent: it is a showcase for future fuel-cell technology. The H₂O is an electric vehicle fitted with batteries and an auxiliary power unit in the form of a fuel cell, the function of which is to provide a permanent supply of electrical energy to power emergency equipment.

To operate, a fuel cell requires hydrogen and oxygen. Oxygen is drawn from the air or from a bottle, whereas the hydrogen must be produced on board when needed. The H₂O, notionally, retains all its functions when in an anaerobic (without oxygen) environment – for instance, when there is a fire in a tunnel or an underground car park. In this case the oxygen necessary for the fuel cell is supplied from two bottles carried on the vehicle.

At the rear of the two-person passenger compartment are a water tank, a telescopic ladder across the top, various storage compartments and an assortment of sockets and connections, and blue flashing beacons – all vital for tackling a blaze.

At the front the large air intake, for cooling the electric motor and brakes, is prominent. The headlight design is rectilinear and shares its style with those on last year's sporty RC concepts.

The interior displays a modern, functional dashboard that groups together various controls. As one would expect, it is red, and it has a few metal finishing touches that are also found around the side vents. At the top are a touch screen, a telephone and a GPS system. In front of the passenger is a second screen linked to a PC that can display maps of large buildings.

Clearly, the H₂O's elaborate premise would be of little practical use in a major emergency, and firefighters would take a dim view of such frivolity. Still, some promise lies behind this cartoonish device.

Design	Gerard Welter
Front tyres	215/45R18
Rear tyres	215/45R18
Length	4294 mm (169.1 in.)
Width	1689 mm (66.5 in.)
Height	1679 mm (66.1 in.)
Wheelbase	2690 mm (105.9 in.)
Track front/rear	1566/1566 mm (61.7/61.7 in.)
Kerb weight	1700 kg (3748 lb.)

Peugeot Sésame

Design	Gerard Welter
Engine	1.6 in-line 4
Power	80 kW (107 bhp)
Gearbox	5-speed manual
Front suspension	MacPherson strut
Brakes front/rear	Discs/discs
Front tyres	R17
Rear tyres	R17
Length	3700 mm (145.7 in.)
Width	1670 mm (65.8 in.)
Height	1630 mm (64.2 in.)
Wheelbase	2310 mm (90.9 in.)

Peugeot has enjoyed massive success in the past decade with B-segment models such as the 205 and 206, yet it is smart enough to keep researching potential new ideas for the market. The Sésame is a mono-volume design for small-car buyers in the sub-B market segment who might like the advantages that sliding doors offer. Mercedes-Benz has already realized a market exists here, and recently launched the Vaneo, based on the A-Class. But the Sésame is smaller than the Vaneo and has just a single sliding door to provide good access to the four individual seats.

The frontal design is similar to that of most recent Peugeot cars. From the edge of the ultra-short bonnet, the windscreen leads to a glass roof complete with two metal roof bars at the edges. Between the bumpers, running around the wheel arches and along the sills, is a small step to give an athletic quality and a solid, defining edge.

To assist its action, the passenger door is guided by a pad sliding on a visible rail sited on the rear wing. This then joins a feature line that continues around the rear of the car, stopping at the spare-wheel cover. At the corners this rail separates the top part, shaped by the side window which connects with the rear screen, and the base, giving the rear-light cluster its own unique shape.

The Sésame is painted in a simple luminous yellow with a matching interior of yellow and dark blue. Depending on Peugeot's courage in probing new markets – not something the company is renowned for – the Sésame could be very close to its upcoming sub-B production offering.

Renault Ellypse

Design	Patrick le Quément
Engine	1.2 in-line 4 turbo-diesel
Power	72 kW (97 bhp)
Torque	200 Nm (147 lb. ft.)
Gearbox	5-speed robotized/shift-by-wire
Installation	Front-engined/front-wheel drive
Front tyres	19 in. PAX run-flat system
Rear tyres	19 in. PAX run-flat system
Length	3930 mm (154.7 in.)
Width	1770 mm (69.7 in.)
Height	1520 mm (59.9 in.)
Wheelbase	2610 mm (102.8 in.)
Track front/rear	1595/1595 mm (62.8/62.8 in.)
Kerb weight	980 kg (2160 lb.)
Fuel consumption	3.2 ltr/100 km (88.3 mpg)
CO_2 emissions	85 g/km

The excitement of the Paris Motor Show every two years is always partly due to the suspense of seeing what Renault will reveal. And in 2002 it was the Ellypse concept, a monospace with softly structured modern surfaces, two-tone colouring – blue-tinted almond for the front and rear, and ultra-light blue for the doors – and a face and tail design that sit comfortably with Renault's brand identity. Its horizontal proportions and similar feature lines, together with bold arcs, give it a friendly feel.

While the left side is equipped with traditional doors, the right has an innovative two-way opening system. There is no B-pillar: the rear door opens either as a classic swing door, to give direct access to the rear seats, or tilts from front to back.

The interior is minimalist, with the shape of the dashboard leading on to the floor to create a gentle wave that supports the four seats. The dashboard also adopts a streamlined design, with two central displays neatly summarizing essential data. One shows driving information, while the other displays 'passenger information' and can be folded away.

The Ellypse's interior utilizes the 'Touch Design' concept first seen on the Talisman concept in 2001. Touch Design is basically a series of simple shapes and 'technology hubs' grouping the different driver functions. The idea is to make controls and functions intuitively easy and pleasant to use.

In addition, the Ellypse explores new ways to make components from recycled materials, particularly for the soundproofing, which is made mainly from cotton from old clothing and polyester fibres from pre-sorted plastic bottles and packaging.

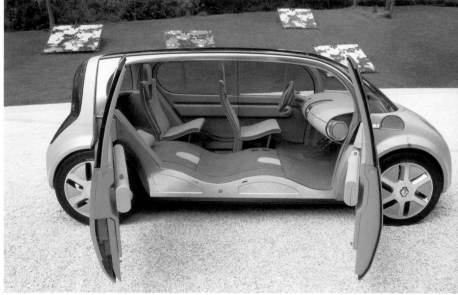

Saab 9-3

Design	Michael Mauer
Engine	2.0 in-line 4 (2.1 in-line 4 diesel also offered)
Power	155 kW (208 bhp) @ 5300 rpm
Torque	300 Nm (221 lb. ft.) @ 2500 rpm
Gearbox	6-speed manual
Installation	Front-engined/front-wheel drive
Front suspension	MacPherson strut
Rear suspension	Independent 4-link
Brakes front/rear	Discs/discs, ABS, TCS, BA
Length	4635 mm (182.5 in.)
Width	1762 mm (69.4 in.)
Height	1466 mm (57.7 in.)
Wheelbase	2675 mm (105.3 in.)
Track front/rear	1524/1506 mm (60/59.3 in.)
Kerb weight	1500 kg (3307 lb.)
0–100 km/h (62 mph)	7.5 sec
Top speed	235 km/h (146 mph)
Fuel consumption	8.5 ltr/100 km (33.2 mpg)

Saab's new 9-3 is a sporty, coupé-like executive saloon with the sort of look of strength we have come to expect from the Swedish brand. The new car is 50 mm (2 in.) wider and about 75 mm (3 in.) longer than the previous 9-3, which gives it a more purposeful stance. Aerodynamics appear to have had a high development priority: the shape at the front is gently raked, leading to the wrap-around windscreen and then across the roof to the tapered rear screen and on to the boot spoiler.

Saab's design DNA reassuringly runs throughout, examples being the wrap-around windscreen, the teardrop form of the side windows with a disguised B-pillar, the wedge shape and the high waistline culminating in the distinctive 'hockey stick' curve into the C-pillar. A single swage line runs the entire length of the car and, of course, those prominent oval-shaped door handles stand out.

Inside, everything is conventional and classic, with soft curves and functional black and white switches and buttons. The interior comes in either 'light room' (parchment) or 'dark room' (grey) tones, the latter being slightly more sporty because it sets off the aluminium detailing more distinctively.

A detail maybe, but Saab has a widely reported feature on its 9-5 model: its cupholder complete with a highly complex mechanism. For the 9-3, a new one was designed, called the 'Butterfly'. This has two separate movements in different directions, one for the base and one for the retaining hoop (together these are the butterfly's 'wings'), which are both hinged from one arm. A highly geared action makes the wings open and close as the arm moves in and out. At least the feature is something genuinely novel in this otherwise gently evolving model.

Sivax Streetster Kira

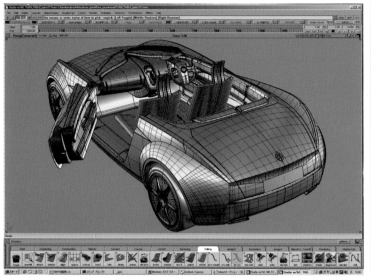 

Sivax, a Japan-based automotive consultancy, brings us this new concept car as a technology and styling demonstrator, taking its inspiration from Tokyo's cosmopolitan street culture and fashion. The name 'Kira' appears in classic Japanese literature and refers to twilled white or plain-coloured silk used for elegant clothing, although it is not clear how this relates to either the exterior or the interior design.

The Kira's design is based on two intertwined wedges, one facing forwards and one rearwards, which meet with intersecting surfaces at the top of the doors. This creates a wedge-shaped feature running in both directions along the side of the car, removing some of the outright sports look that would otherwise be achieved by such an uncompromising open-top sports car.

At the front, the striking headlights form part of the V-shaped feature that runs alongside the bonnet and up to form the waistline wedge. A centre-line spine, more familiar from Renault designs, rises up over the bonnet and splits the windscreen, which is apparently modelled on a pair of sports sunglasses.

The two-plus-two interior combines leather, fabric and metal; the instrument panel uses brass while the floor has pear-skin-finished aluminium, the product of a street-fashion design technique.

The Kira Streetster is purely a concept intended to draw attention to the development capabilities of Sivax. Although not likely to win any design awards, it has a few interesting features, particularly inside, where there is some original thought.

Engine	2.0
Length	3980 mm (156.7 in.)
Width	1800 mm (70.9 in.)
Height	1050 mm (41.3 in.)
Wheelbase	2500 mm (98.4 in.)

Toyota Land Cruiser

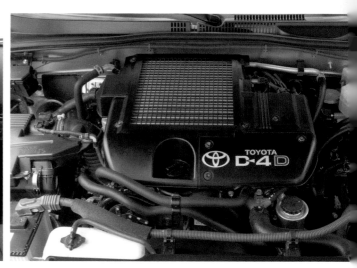

Design	Toyota Motor Corporation's ED2 studio
Engine	4.6 V6 (3.0 in-line 4 diesel also offered)
Power	183 kW (245 bhp) @ 5200 rpm
Torque	382 Nm (282 lb. ft.) @ 3200 rpm
Gearbox	4-speed automatic
Installation	Front-engined/4-wheel drive
Front suspension	Double wishbone
Rear suspension	Independent 4-link
Brakes front/rear	Discs/discs, BA, ABS, EBD
Front tyres	265/65R17
Rear tyres	265/65R17
Length	4365 mm (171.9 in.)
Width	1790 mm (70.5 in.)
Height	1890 mm (74.4 in.)
Wheelbase	2455 mm (96.7 in.)
Track front/rear	1535/1535 mm (60.4 in.)
Kerb weight	1750 kg (3858 lb.)
0–100 km/h (62 mph)	9.5 sec
Top speed	175 km/h (108.8 mph)

The Paris Motor Show in 2002 marked the world debut of the all-new Toyota Land Cruiser. Toyota celebrated fifty years of off-roader manufacture in 2001, a half-century since the first Land Cruiser – a light, all-purpose 4×4 for civilian and military applications – took its bow, and the company's 4×4 output has been an important cornerstone of its prosperity.

The styling of the new Land Cruiser was done in Europe at ED2, Toyota's advanced design facility on the Côte d'Azur, in the south of France. This is the first time that a Land Cruiser exterior has been designed outside Japan and is therefore a measure of Toyota's confidence in its satellite studio. At the front, much of the sense of presence comes from the tough-looking, vertically veined grille, the high bonnet and the bulging body mouldings that protect the car when off road. New technology on the car includes a world first: Hill-start Assist Control, for ascending slippery slopes, and Downhill Assist Control for descending.

The new Land Cruiser retains a traditional frame chassis and separate body construction, unlike the BMW X5 and Range Rover, which now have highly reinforced monocoques to take the suspension loads and provide increased stiffness. They handle almost like saloon cars, but a monocoque has been rejected by Toyota: not for nothing do Toyota 4×4s abound in the background of news reports from the world's most inhospitable corners; the X5 is nowhere to be seen. For the Land Cruiser is aimed at the premium luxury-car segment but at the same time offers unstoppable off-road capability.

Inside, in an attempt to be both luxurious and rugged, a dull, outdated and predominantly plastic interior has been ham-fistedly jazzed up with leather, wood and heavy aluminium details. Too many materials in an uninspiring overall design: quite a contrast to the sophistication of the Porsche Cayenne (see p. 98) or the Volkswagen Touareg (see p. 238).

The Design Process

Profiles of Key Designers

Why Concept Cars?

Concept Car Evolutions

Super-Luxury Cars

Super-Luxury Cars

Left and opposite
Maybach

Super-luxury, or 'super-lux', as the industry has it, means cars with a price of £120,000 and upwards. It's currently a segment largely dominated by such contrasting brands as Rolls-Royce, Bentley, Ferrari and Lamborghini. Despite the effects of global uncertainty, plunging stock markets and the long-gone dot.com bubble, 2003 and 2004 will bring an influx of new super-lux models to choose from, a charge led by the relaunch of some historic brands. The reason for this is the relentless buying-up of historic marques over the past decade. Rolls-Royce, Bentley, Bugatti and Lamborghini have all been acquired by large German car manufacturers such as BMW and Volkswagen, while DaimlerChrysler has raided its own cellars to revive Maybach, a marque acquired in 1960 and last seen gracing a car in 1941.

It's demographic trends that are luring car makers into the super-lux segment. There are more and more rich people and their numbers are predicted to grow strongly over the next decade. In the USA alone, sales of super-lux cars are expected to double by 2010 to about six thousand vehicles a year.

Buying such a car sends a strong message to business associates and friends, but manufacturers who offer them are anxious to avoid the negative stigma these cars can attract. In 1936 the number of people in the USA who could afford to buy a Duesenberg was the same as in 1929, but the company's luck ran out: it went out of business because it wasn't considered appropriate to buy and be seen to own that kind of car during the Depression. Although many potential customers today may be insulated from short-term economic downturns, if the US economy takes a further prolonged dip and this in turn affects the Japanese and European markets, there may be a serious shortage of buyers once more.

Volkswagen is the most aggressive of the large car makers entering the super-lux segment. It now owns Bentley, Bugatti and Lamborghini. It has also extended the reach of its core Volkswagen and Audi brands upwards, and it owns Horch, another historic luxury brand widely rumoured to be a candidate for revival. The all-new Bentley Continental GT goes into production at the original plant in Crewe, England, where both Rolls-Royces and Bentleys have traditionally been built. A number of other Bentley derivatives are expected, which is great news for the average Bentley owner, with his or her typical net worth of £8–12 million.

Technical Glossary

Where the New Models were Launched

Major International Motor Shows 2003–2004

Marques and their Parent Companies

Major International Motor Shows 2003–2004

**Paris Motor Show
(Mondial de l'automobile)**
Sunday 28 September – Monday 13 October 2003
Paris Expo, Paris, France
www.mondialauto.tm.fr

Budapest Motor Show
Tuesday 14 – Sunday 19 October 2003
Budapest Fair Centre, Budapest, Hungary
www.automobil.hungexpo.hu

Prague Auto Show (Autoshow Praha)
Thursday 16 – Sunday 19 October 2003
Prague Exhibition Grounds, Prague,
Czech Republic
www.incheba.cz

Tokyo Motor Show
Saturday 25 October – Wednesday 5 November
2003
Nippon Convention Centre, Chiba City, Tokyo
www.tokyo-motorshow.com

Riyadh Motor Show
Sunday 9 – Friday 14 November 2003
Riyadh Exhibition Centre, Riyadh, Saudi Arabia
www.recexpo.com

Middle East International Motor Show
Thursday 11 – Monday 15 December 2003
Dubai World Trade Centre (DWTC), Dubai,
United Arab Emirates
www.dubaimotorshow.com

Greater LA Auto Show
Friday 2 – Sunday 11 January 2004
Los Angeles Convention Center, Los Angeles, USA
www.laautoshow.com

**North American International Auto Show
(NAIAS)**
Saturday 10 – Monday 19 January 2004
Cobo Center, Detroit, USA
www.naias.com

Brussels International Motor Show
Thursday 15 – Sunday 25 January 2004
Brussels Expo, Brussels, Belgium
www.febiac.be/motorshows

Chicago Auto Show
Friday 6 – Sunday 15 February 2004
McCormick Place South, Chicago, USA
www.chicagoautoshow.com

Canadian International Auto Show
Friday 13 – Sunday 22 February 2004
Metro Toronto Convention Centre and SkyDome,
Toronto, Canada
www.autoshow.ca

Melbourne International Motor Show
Thursday 26 February – Monday 8 March 2004
Melbourne Exhibition Centre, Melbourne,
Australia
www.motorshow.com.au

74th Geneva International Motor Show
Thursday 4 – Sunday 14 March 2004
Palexpo, Geneva, Switzerland
www.salon-auto.ch

Tallinn Motor Show (Motorex 2004)
Wednesday 24 – Sunday 28 March 2004
The Estonian Fairs Centre
Tallinn, Estonia
www.fair.ee

New York International Auto Show
Friday 9 – Sunday 18 April 2004
The Jacob Javits Convention Center,
New York, USA
www.autoshowny.com

British International Motor Show
Tuesday 25 May – Sunday 6 June 2004
National Exhibition Centre (NEC), Birmingham, UK
www.motorshow.co.uk

**Paris Motor Show
(Mondial de l'automobile)**
Saturday 2 – Sunday 17 October 2004
Paris Expo, Paris, France
www.mondialauto.tm.fr

Marques and their Parent Companies

Hundreds of separate car-making companies have consolidated over the past decade into ten groups: General Motors, Ford, DaimlerChrysler, VW, Toyota, Peugeot, Renault, BMW, Honda and Hyundai. These account for at least nine of every ten cars produced globally today.

The remaining independent marques either produce specialist models, offer niche design and engineering services or tend to be at risk because of their lack of economies of scale.

The general global improvement in vehicle quality means manufacturers would be wise to continue to rely heavily on design to differentiate their brands.

BMW
BMW
Mini
Riley*
Rolls-Royce
Triumph*

DaimlerChrysler
Chrysler
De Soto*
Dodge
Hudson*
Imperial*
Jeep
Maybach
Mercedes-Benz
Mitsubishi
Nash*
Plymouth*
Smart

Fiat Auto
Abarth*
Alfa Romeo
Autobianchi*
Ferrari
Fiat
Innocenti*
Lancia
Maserati

Ford
Aston Martin
Daimler
Ford
Jaguar
Lagonda*
Land Rover
Lincoln

Mazda
Mercury
Range Rover
Th!nk*
Volvo

General Motors
Buick
Cadillac
Chevrolet
Daewoo
GMC
Holden
Hummer
Isuzu
Oldsmobile*
Opel
Pontiac
Saab
Saturn
Subaru
Suzuki
Vauxhall

Honda
Acura
Honda

Hyundai
Asia Motors
Hyundai
Kia

MG Rover
Austin*
MG
Morris*
Rover
Wolseley*

Peugeot
Citroën
Hillman*
Humber*
Panhard*
Peugeot
Simca*
Singer*
Sunbeam*
Talbot*

Proton
Lotus
Proton

Renault
Alpine*
Dacia
Datsun*
Infiniti
Nissan
Renault
Renault Sport

Toyota
Daihatsu
Lexus
Toyota

VW
Audi
Auto Union*
Bentley
Bugatti
Cosworth
DKW*
Horch*
Lamborghini
NSU*

Seat
Skoda
Volkswagen
Wanderer*

Independent marques
Austin-Healey*
Bertone
Bristol
Caterham
Fioravanti
Heuliez
Invicta
Irmscher
Italdesign
Lada
Koenigsegg
Matra
Mitsuoka
Morgan
Pininfarina
Porsche
Rinspeed
Sivax
SsangYong
Tata
TVR
Venturi
Westfield
Zagato

* Dormant marques

Picture Credits

The illustrations in this book have been reproduced with the kind permission of the following manufacturers:

Alfa Romeo
Aston Martin Lagonda
Audi AG
Bentley Motors
Bertone SpA
BMW AG
Buick
Cadillac
Chevrolet
Citroën
GM Daewoo Motors
DaimlerChrysler
Dodge
Ferrari SpA
Fiat Auto
Ford Motor Company
General Motors Corporation

Honda Motor Co
Hyundai Car UK
Invicta Car Company
Italdesign
Jaguar Cars
Kia Motors Corporation
Automobili Lamborghini
Lancia
Lincoln
Matra
Mazda Motors
Mercury
MG Rover Group
Mitsubishi Motors Corporation
Nissan Motors
Opel AG
Peugeot SA

Pininfarina SpA
Pontiac
Porsche AG
Renault SA
Rinspeed
Rolls-Royce Motor Cars
Saab Automobile AB
Saturn
Seat SA
Sivax
Subaru
Suzuki Motor Corporation
Toyota Motor Corporation
TVR Engineering
Volkswagen AG
Volvo Car Corporation

Acknowledgements

I would firstly like to thank everyone at Merrell Publishers who has helped to ensure that *The Car Design Yearbook 1* was such a fabulous success, and for their commitment to building on that success for future editions. I would especially like to thank Anthea Snow, Kate Ward, Emily Sanders and Emilie Nangle.

Thanks are also due to the manufacturers' press offices, which have been extremely supportive once again in providing technical and photographic material. I would like to thank Vicky Gallagher for her continual support, and Peter Newbury for his assistance in researching the technical specifications. I must also thank Giles Chapman and Richard Dawes for their professional editorial support, and Alistair Layzell for his success in the public-relations campaign.

Stephen Newbury
Henley-on-Thames, Oxfordshire
2003